FURNACE OF CREATION
CRADLE OF DESTRUCTION

Frontispiece to Thomas Burnet's *The Sacred Theory of the Earth,*
published in 1681.

According to the model proposed by Burnet, the antediluvian earth was a smooth ovoid, or egg, with a uniform texture. When the Flood came, sections of a weakened crust cracked open, exposing the water layer beneath. Some of the cracked crust fell into the interior and the Flood waters burst onto the surface. In this process, mountains were upturned fragments of the early crust left behind after the Deluge, and oceans were holes left where the crust broke up. This was one of the first theories on the formation of the earth's features that attempted to reconcile the views of the Church and Science.

rped

FURNACE OF CREATION

CRADLE OF DESTRUCTION

A Journey to the Birthplace of Earthquakes, Volcanoes, and Tsunamis

ROY CHESTER, PH.D.

AMACOM

American Management Association
New York * Atlanta * Brussels * Chicago * Mexico City * San Francisco
Shanghai * Tokyo * Toronto * Washington, D.C.

Special discounts on bulk quantities of AMACOM books are available to corporations, professional associations, and other organizations. For details, contact Special Sales Department, AMACOM, a division of American Management Association, 1601 Broadway, New York, NY 10019.
Tel.: 212-903-8316 Fax: 212-903-8083
E-mail: specialsls@amanet.org
Website: www.amacombooks.org/go/specialsales
To view all AMACOM titles go to: www.amacombooks.org

This publication is designed to provide accurate and authoritative information in regard to the subject matter covered. It is sold with the understanding that the publisher is not engaged in rendering legal, accounting, or other professional service. If legal advice or other expert assistance is required, the services of a competent professional person should be sought.

Library of Congress Cataloging-in-Publication Data

Chester, R. (Roy), 1936–
    Furnace of creation, cradle of destruction : a journey to the birthplace of earthquakes, volcanoes, and tsunamis / Roy Chester.
        p.   cm.
    Includes index.
    ISBN-13: 978-0-8144-0920-6
    ISBN-10: 0-8144-0920-2
    1. Plate tectonics.   2. Continental drift.   I. Title.

QE511.4.C44   2008
551.1'36—dc22   2008012702

Printing number

10   9   8   7   6   5   4   3   2   1

To Oliver, Edward, and Henry

"The next generation"

CONTENTS

CONTENTS

ACKNOWLEDGMENTS

I would like to express my personal thanks to my long-suffering wife, Alison. Writing is a lonely business and she has, as usual, stoically put up with my "absences" from normal family life as I have shut myself away for periods of time that seem to increase exponentially as deadlines approach.

I am grateful to those individuals, and organizations, who kindly gave permission for the reproduction of figures, and I have been happy to credit each one at the appropriate place in the text. My particular thanks go to Professor Steven Nelson for generously allowing me to use so many of his diagrams, and to the United States Geological Survey for making available the range of figures from its collection. I would also like to acknowledge my debt to Jennifer Holder, whose editorial input made such a great difference to the book.

FURNACE OF CREATION
CRADLE OF DESTRUCTION

INTRODUCTION
THE TIME BEFORE SCIENCE

If we needed an event to remind us of the great danger that could arise from natural disasters, then the Indian Ocean tsunami of 2004 played the role well. The tragedy of the Asian earthquake, and the Indian Ocean tsunami that followed it, unfolded over the period of Christmas 2004. Viewers watched in horror as television screens across the world broadcast stark images of death and destruction as a tsunami struck the coastlines of the Indian Ocean. This was real. This was disaster brought into the living room.

The response from the public was immediate and unstinting. But there was something else here as well. People had witnessed the raw power of nature at its destructive worst, and there was a thirst for understanding how it had happened. I had retired from my university post at that time, but I was asked to give a series of public lectures on volcanoes, earthquakes, and tsunamis—that great trinity of natural disasters. During these lectures, I found audiences were fascinated by the fact that there was one single underlying thread that controlled the way the surface of our planet had evolved; a thread that underpinned the way volcanoes, earthquakes, and tsunamis were generated. That thread was the theory of plate tectonics.

Using this theory, I could explain to audiences the causes of the trinity of natural disasters in a scientific manner. But this raised a question with me: How would volcanoes, earthquakes, and tsunamis have appeared to people in the past, before the advent of modern theories about plate tectonics or, even more intriguing, in the *time before science*? How would ancient cultures view destructive visitations in the shape of these natural phenomena that threatened to destroy their world? And how did our thinking evolve as we moved toward the system of knowledge in place today?

To try to answer questions such as these, I have gone back in

1

time. First to the early cultures, where notions of the natural world were deeply wrapped in myth and legend. Then to a period in which scientific thought was starting to emerge, but where ideas were constrained by religious beliefs. Finally, to the time when restrictions were finally thrown off and the only constraint to scientific progress was the limitation of human ingenuity.

▲

A central theme in understanding plate tectonics is the notion that the continents have somehow wandered about the surface of the earth. Moving continents is not a new idea. It first emerged during the great "age of exploration," when the first reliable maps of the world began to appear. As early as 1596, the Dutch mapmaker Abraham Ortelius, in his work *Thesaurus Geographic,* put forward the idea that that the Americas had broken away from Europe. In 1620, Francis Bacon, the English philosopher, court politician, and "father of deductive reasoning," remarked on how the west coast of Africa and the east coast of South America appeared to fit together. So close was the fit that he suggested the continents of America and Africa had, in fact, once been joined, and later it was suggested that they had been separated by the flood—perhaps the first attempt to explain the mechanism behind the movement of the continents.

Toward the end of the twentieth century, two oceanographers diving in the deep-sea submersible *Alvin* on the Galapagos Ridge in the Pacific Ocean came upon one of the most astonishing sights in the history of natural science. They had discovered a staggeringly new biological community of mainly unknown animal species: massive clumps of large red tube-dwelling worms, fields of giant clams and mussels and blind crabs, all living around hot springs emerging from the seabed on an underwater mountain range.

More than 300 years separate the emergence of the earliest suggestions that the continents had moved and the discovery of the Galapagos hot springs, but both developments may be thought of as crucial stages along the timeline of a great scientific revolution during which the concept of continental drift led to the theory of seafloor spreading and finally to our current understanding of plate tectonics.

For earth science, the theory of plate tectonics was as important as Darwin's *Origin of Species* was for biology and the Theory of Everything will be for physics. The reason is that plate tectonics offers a unified theory to explain the way the earth has evolved by identifying the processes that have governed the way the surface features of the planet have developed—processes that also control volcanoes, earthquakes, and tsunamis. In fact, so far-reaching were the implications of the plate tectonics theory that the oft-overused expression "the textbooks had to be rewritten" was, in this case, literally true.

Often science only advances when currently held wisdom is challenged, and the road to plate tectonics is littered with some of the greatest controversies in the history of geology. Driven forward by wave after wave of new evidence, however, the revolution gathered momentum until eventually it became unstoppable. The new evidence came from many sources and brought together scientists in the fields of physics, geophysics, chemistry, biology, geology, and the relatively young discipline of oceanography. As the various lines of evidence began to gel, it became apparent that unravelling the full story of plate tectonics was like solving a massive scientific jigsaw puzzle.

The development of plate tectonics is one of the greatest stories in the forward march of modern science. By using as our central theme the changes in the way humankind has viewed volcanoes, earthquakes, and tsunamis, we will follow the journey that led to plate tectonics from its origins in myth and legend to the science of modern times.

That is the purpose of the book—to portray the progress of human understanding from ancient mythmaking to scientific enlightenment.

1
SETTING THE SCENE

THE TRINITY OF NATURAL DISASTERS

Natural disasters come in many guises. Volcanoes and earthquakes are the result of shifts within the earth's crust and the molten material below it, and both can set off tsunamis—movements within the earth's oceans. Volcanoes, earthquakes, and tsunamis are, when taken together, what might be termed a "trinity of natural disasters."

Over geological time volcanoes, earthquakes, and tsunamis have struck the earth with varying degrees of ferocity, and from the age of the most ancient cultures, humankind has attempted to explain these visitations of nature's fury.

MYTHS AND LEGENDS

One of the characteristics that distinguished early humans from other animals was their need to understand the world around them, and this applied especially to natural phenomena that brought danger to the environment they lived in. All three of the great trinity of natural disasters carry a grave risk of danger, and together they are among the most terrifying phenomena ever to have been visited on the earth. Little wonder, then, that volcanoes, earthquakes, and tsunamis became the stuff of myth and legend.

In an age of scientific reason, these myths may seem ridiculous, but they should be seen for what they were at the time—attempts to explain nature within a context that the people of ancient societies could understand. Later, the myths and legends gave way to more rational thought, but it was to be a long road spanning many centuries before a rigorous scientific protocol was developed that could fully explain these natural disasters.

We will travel this road, beginning with the emergence of the myths and legends and ending at the flowering of the theory of plate tectonics. The terms *myths* and *legends* are difficult to distinguish from each other and are considered interchangeable here.

▲

What are volcanoes, earthquakes, and tsunamis? Later, we will define them in the context of modern science, but for now it is interesting to see them as ancient peoples might have seen them—not as rational events, but as acts of the gods.

Volcanoes: furnaces that throw fire and molten rock out of the depths of the earth.

Earthquakes: devils that make the ground twist and tremble and buckle.

Tsunamis: giant destructive waves that appear suddenly out of the sea to hammer the land.

To the people living in early societies, these were great forces that acted on the world they lived in—and were rich and fertile grounds for the shaman and the witch doctor. And in a climate in which superstition was rife, different cultures spawned their own individual myths and legends in an attempt to explain these visitations of nature.

Volcanoes, with their fire and molten rock, have always inspired fear and generated myths. One colorful belief among the peoples of Kamchatka, a Russian peninsula with several large volcanoes, was that after the world was created by the god Kutkh, the Great Raven, the first men were given a beautiful woman. When the men died they were turned into mountains that became volcanoes when their hearts burned with love for the woman. In another myth, also from Kamchatka, the natives believed that *gomuls*, spirits that live in mountains, hunted for whales that they cooked in volcanoes, causing whale fat to run down the sides of the peaks. In Chile, the natives also believed that a great whale lived inside volcanoes.

This association of volcanoes and giant animals is common, and one Japanese myth has a massive spider dwelling within the earth and causing volcanoes to erupt. Volcanoes have always played a central role in the mythology of Japan—a land with more than a hundred active volcanoes. The most famous of all the Japanese volcanoes is the sacred Mount Fuji (*Fujiyama*—the Never Dying Mountain). This symmetrical cone-shaped peak is a national symbol, and legend has it that it is the home of a Fire Spirit. To the natives of New Zealand, volcanoes are also the work of Fire Demons that can cause the land to burst into flame.

Some of the most colorful volcanic legends come from the Hawaiian island chain. Here, the legends describe how Pele, the beautiful goddess of volcanoes, caused eruptions when she became angry, which apparently was quite often, by scratching the surface of the ground with a magic stick (*Pa'oe'*). She could also bring about earthquakes by stamping her feet. But not all volcanic myths involved violence, and my own favorite is the Aztec legend in which Popocatepetl (the Smoking Mountain) and an adjacent peak (the White Lady) were lovers that could not be parted.

Battles of the gods were common in volcanic myths. According to the Klamath Native Americans, the great eruption of Mount Mazama on the western U.S. mainland was caused by a battle between Llao (god of the underworld) and Skell (god of the sky). Volcanoes, with their dramatic explosions of fire and clouds of ash hurled into the air, must have appeared to ancient humans as if the gods were ripping the planet apart. Not surprising, then, that volcanoes became associated with the Fires of Hell—as in the legends of Iceland, where volcanoes such as Hekla are considered a gateway into the underworld. Volcanic disasters have also played a part in stories of creation and destruction, one of the most famous being the destruction of Atlantis by a violent volcanic eruption that split the land apart.

When Vesuvius erupted massively in AD 79, its impact on society has to be seen in terms of the then-current Greek/Roman mythology, which held the gods responsible for natural disasters. In this instance, the Greek god of fire, Hephaestus (later Vulcan, the blacksmith to the Roman gods), lived under mountains in his forge where he made thunderbolts. Indeed, the word *volcano* comes from the island of Vulcano in the Mediterranean Sea, where the inhabitants originally believed that they were living on the chimney of the forge of Vulcan. It was against this background that Vesuvius erupted with such fatal consequences in AD 79. In the hours of daylight, but under a blackened sky, sheets of fire and clouds of ash and sulfurous vapor were flung from the volcano and the ground shook. Men, women, and children cried out in panic, believing the gods had deserted them (Panel 1).

PANEL 1
MOUNT VESUVIUS*
(ITALIAN MAINLAND, AD 79)

"The death of Pompeii"

A convergent plate boundary, subduction zone stratovolcano with a height of around 1,200 meters, Vesuvius has erupted many times. The most violent eruption, which occurred in AD 79, was of the Plinian type—that is, it was a spectacular explosive eruption of viscous lava that sent a cloud of volcanic ash and gases to a height of several miles, when it then spread out horizontally.

Death toll—unknown—although estimates range from 3,500 up to 20,000.

Principal cause of death—thermal shock from ash falls and pyroclastic flows.

▲

The destruction of the Roman towns of Pompeii and Herculaneum by Vesuvius in AD 79 has become the iconic volcanic disaster in the western psyche. Vesuvius, which is located in the beautiful landscape of the Bay of Naples, is part of the Campanian volcanic arc and is believed to be one of the most dangerous volcanoes in the world.

The AD 79 Vesuvius eruption is thought to have lasted for around twenty hours and to have released 4 cubic kilometers of ash and other pyroclastic material over an area to the south and southeast of the mountain.

Apparently, there were some portents of the disaster in the mid-summer of AD 79; for example, springs and wells in the vicinity dried up, and a number of small earthquakes occurred. The Romans were used to natural disturbances in the area, and earthquakes were not uncommon. However, there was no suggestion that they were linked to volcanic explosions, and the local population was totally unprepared for the catastrophic eruption of Vesuvius that took place in the afternoon of August 24.

When the mountain exploded, a great cloud of ash, lava, and gases, including sulfur dioxide, was hurled into the air. As the cloud dispersed, a firestorm of volcanic ejecta and toxic gases engulfed the area around Vesuvius, blanketing buildings and burying the population of Pompeii and Herculaneum. Such was the extent of the damage that the towns were buried under ejecta (mainly ash) up to 20 meters deep and were not, in fact, rediscovered until the eighteenth century.

The eruption was witnessed by Pliny the Younger from Misenum, across the bay from Vesuvius. In the early afternoon he saw a dense cloud, shaped like a tree trunk, rise above the mountain, then split into two branches. In some places the cloud was white, in others it was gray and dirty. Later, part of the cloud flowed down the sides of the mountain, blanketing everything in its path. This was what is now called a "pyroclastic flow." As the cloud spread out, the sun was cut off and day turned to night.

When Pompeii was rediscovered, a majority of the bodies (around 60 percent) were found buried under the pyroclastic surge deposits; the people probably died from thermal shock and suffocation from the pyroclastic cloud. Pompeii suffered, too, from volcanic ash and tephra falls, and most of the remaining casualties were found beneath ash deposits, often trapped inside buildings that were buried under as much as 20 meters of ash. Herculaneum was spared from the worst effects of falling tephra by the wind direction, but it was buried under pyroclastic deposits.

Today, Vesuvius is a national park, with more than half a million people living within the Red Zone (Zona Rossa) around the volcano, and the authorities have an evacuation plan should it be necessary to move large numbers of people quickly. However, the plan is thought by some experts to be flawed, because it would require a longer warning period than the scientists monitoring the volcano could provide. Critics of the evacuation plan also believe the Red Zone should be extended—perhaps even to include the city of Naples.

*The classification of the volcanoes, earthquakes, and tsunamis in the panels is based on post–plate tectonic thinking, and the terms used are described in the text.

Although myths abounded, there were a number of attempts by the Greco-Romans to offer quasi-scientific explanations for natural disasters. One theory, for example, proposed that there was a great river of fire flowing through the interior of the earth, supplying all the volcanoes on the surface.

Myths are also held by more modern native people. According to a New Zealand Maori legend woven around the 1886 eruption of the Tarawera volcano, the morals of the local people had fallen under the influence of white settlers, and a man-eating demon, Tamaohoi, was summoned from sleep to punish the sinners. In response, the demon spewed out as a volcano, killing over a hundred people and destroying three villages.

Although science continued to forge forward, there are many other examples of native myths surrounding volcanoes. The great volcanic eruptions of Tambora in 1815 (Panel 2) and Krakatau in 1883 (Panel 3) were prime examples of volcanic activity occurring in regions rich in superstition. Here, the natives of Indonesia believed that a great snake, Hontobogo, lived below the earth's surface. When the snake became restless and moved, the earth shook and spat out fire.

PANEL 2
VOLCANOES OF THE RING OF FIRE, INDONESIA*

"The year without a summer"

Lying between the Pacific Ocean *"Ring of Fire"* and the Alpide volcanic belt, Indonesia, which consists of more than 17,500 islands, is the volcanic capital of the world. The estimated total number of active volcanoes in the area ranges between 75 and 130, with approximately 75 percent of them lying in the Sunda Arc system. The island arc was formed as a consequence of the collision of the Indo-Australian and the Eurasian tectonic plates. This event resulted in the subduction of the largely oceanic (Indo-Australian) plate under the continental plate (Eurasian), and led to the formation of one of the most intense volcanic and earthquake activity zones in the world.

Two volcanoes, Mount Tambora and Krakatau, can be considered to be characteristic of the Indonesia region. Both are "convergent plate boundary, subduction zone" volcanoes with pyroclastic cloud eruptions. Together, Tambora and Krakatau occupy fist and second positions, respectively, on the basis of death tolls, in the table of *Deadliest Volcanic Eruptions Since 1500 AD*. Both volcanoes have been responsible for great loss of life, and both have had significant effects on the global environment.

MOUNT TAMBORA
(SUMBAWA ISLAND—INDONESIA, 1815)

Volcanic Explosivity Index—~7.

Estimated death toll—71, 000 to 92,000.

Principal cause of death—crop destruction leading to starvation.

▲

Mount Tambora, which is a stratovolcano with a height of 2,800 meters, lies above a subduction zone on the Sunda Arc. During its history, it developed a volcanic cone with a single central vent.

The 1815 eruption occurred over a period of several days in April. It was the most violent since that of Lake Taupe in AD 18, and the blast of the major eruption was heard as far away as 2,500 kilometers. Large quantities of sulfur dioxide were released into the air, and the column of ash and smoke discharged reached a height of 45 kilometers, with ash layers of one centimeter reported as far away as 1,000 kilometers. After the initial eruption, several pyroclastic flows developed, some of which reached the ocean, causing further explosions.

The total death toll was estimated to be as high as 90,000. Some 10,000 people are thought to have died from direct effects of the eruption, such as pyroclastic flows. This fact alone would have put Tambora in the big league of volcanic eruptions. In addition, it has been estimated that as many as 80,000 people perished from famine, starvation, and disease resulting from the volcanic activity. Many of the deaths from these secondary volcanic hazards were in and around Indonesia, but the effects of the eruption were felt on a global scale.

The eruption of Tambora is of special interest because it provides an example of how volcanic activity can affect climate, and cause death, on a global scale. During the eruption, volcanic dust and ash was lifted into the stratosphere and traveled around the earth several times. As they did so, they blocked out sunlight and so reduced the amount of solar radiation reaching the surface of the planet. Estimates of the average global temperature drop range from 0.4°C to 0.7°C, but values as high as 3°C have been suggested. Whatever the actual figure, this temperature drop is thought to have been responsible for a whole raft of disastrous climatic effects, including severe winds and rainstorms and a shift in the seasons. In the United States, the effects included a population migration from the colder climates of New England to the warmer Midwest. In Europe, there was flooding, and in northern Europe especially crop failure led to civil unrest together with widespread famine and death. A typhus epidemic has also been attributed to the general conditions arising from the change in the climate and famine.

Other dramatic, but less severe, effects of the eruption of Mount Tambora included vivid-colored sunsets witnessed all over the world, reports of red and brown snow falling in Europe, and the formation of a sulfate aerosol veil, from the reaction of sulfur dioxide with water vapor in the atmosphere.

In some places the climate was turned on its head, with frost and snow in summer. In fact, so acute were the climatic anomalies resulting from the Tambora eruption that the following year, 1816, became known as "the year without a summer."

*The classification of the volcanoes, earthquakes, and tsunamis in the panels is based on post–plate tectonic thinking, and the terms used are described in the text.

PANEL 3
THE KRAKATAU VOLCANO AND TSUNAMI*
(INDONESIA, 1883)

"The Loudest Sound in History"

Tsunami classification—volcano-generated.

Location of volcano—Sunda Strait.

Tectonic setting—The area lies in the Malay Archipelago and sits in a tectonically active region on the boundaries of three tectonic plates: the Pacific, the Eurasian, and the Indo-Australian. The Indonesian island arc system is a result of the subduction of the Indo-Australian plate under the Eurasian plate.

Overall death toll—~36,000.

Principal cause of death—the tsunami.

▲

Krakatau, a volcanic island lying between Java and Sumatra, had erupted before during recorded history, and prior to the 1883 eruption the last one had been in 1680. At one stage in its history, the top of the mountain had been blown away forming a caldera, and three cones projected as islands above sea level, the highest reaching around 800 meters.

On May 20, 1883, one of the cones began to erupt clouds of dust and ash into the air to a height of 10 kilometers. For the next two months the volcano continued to rumble and spew out black clouds, and Krakatau became a kind of tourist attraction with sightseers charting ships to view the spectacle. Then the preliminaries culminated when, on August 26, 1883, Krakatau gave vent over two days to one of the greatest volcanic explosions in recorded history.

The eruption of Krakatau had a Volcanic Explosivity

Index (VEI) value of 6, more than 10,000 times greater than the atomic bomb dropped on Hiroshima in the Second World War. The noise of the explosion was heard up to almost 5,000 kilometers away and has been called the "loudest sound in history."

Shock waves reverberated through the atmosphere for days afterward, and the whole event was made more sinister because ash clouds brought darkness to the surrounding region, terrifying the inhabitants. To give some idea of the scale of the ash clouds, material 60 meters deep was deposited in the vicinity of the eruption, and ash falls covered an area of approximately 60,000 square kilometers.

The activity ended on August 27 in a massive climax in which a series of eruptions tore the island apart, much of it disappearing beneath the sea. In addition, giant pyroclastic flows poured into the water, initiating mountainous tsunamis. The total death toll from the eruption of Krakatau has been estimated to be 36,000. But few people died in the immediate area of the eruption because Krakatau was an uninhabited island. Most fatalities occurred as the tsunamis, some over 40 meters high, hit the coastal areas of the islands and the Indonesian mainland, with as many as 30,000 deaths in the coastal towns of Java and Sumatra. Around 4,500 people perished in other ways; for example, in pyroclastic flows or as a consequence of debris falling on them from the air. Incredibly, the pyroclastic flows appeared to have traveled considerable distances, as much as 40 kilometers, across open water.

In addition to the terrible loss of life, Krakatau had a number of global effects, mainly due to the volcanic ash that was blown into the stratosphere and traveled several times around the earth. Furthermore, large quantities of sulfur dioxide were emitted during the eruptions, leading to the production of sulfuric acid aerosols. These various aerosols caused vivid-colored sunrises and sunsets in the sky and cut down solar radiation.

*The classification of the volcanoes, earthquakes, and tsunamis in the panels is based on post–plate tectonic thinking, and the terms used are described in the text.

Earthquakes and tsunamis have also inspired their fair share of myths and legends. In one fanciful Indian tale, the world rests on the head of an elephant and when the animal moves, an earthquake results. In a Greek myth, the earth is thought to float on a great sea that can be agitated to form earthquakes, and in another it is believed that winds are trapped in caverns under the ground and the movements they cause as they try to escape bring about earthquakes.

In a North American myth, earthquakes result when a turtle carrying the earth on its back moves, and in Japanese mythology the thrashing of a giant catfish leads to earthquakes. This theme of earthquakes resulting from the actions of giant creatures recurs in different cultures. In my view, one of the most exotic of all the creature myths originates in Africa and tells how a fish carries a stone on its back on which stands a cow holding the earth on one of her horns. For a respite, the cow throws the globe from one horn to another, making the earth tremble.

In Nordic folklore, the god Loki is punished for killing his brother by having a serpent drip poison on him, and the earth moves as he struggles to avoid it reaching him. In a much more gentle setting, a charming New Zealand legend has Mother Earth carrying a child who causes earthquakes as he moves about in her womb.

The Chinese had several myths associated with the causes of earthquakes. In one, dragons sleeping under mountains caused the earth to move as they stirred. In another, the earth was thought to rest on the back of a giant frog that sets off earthquakes when it shivers. China has a long history of earthquakes, and one of the earliest seismic events for which there are records occurred there. This was the great Shaanxi (Shansi) earthquake of 1556, which was also one of the deadliest of all time (Panel 4).

The terror felt by the people of the Shansi area, so many of whom were trapped in collapsing caves, must have been almost beyond understanding as they felt the spirits unleashing their fury on them. This was in keeping with the popular Chinese belief that natural phenomena resulted from the disturbance of mythical cosmic forces. But there was a paradox here, because Chinese scholars had been involved in the study of earthquakes for many years. In

PANEL 4
THE SHAANXI (SHANSI) EARTHQUAKE*
(SHAANXI, CHINA, 1556)

"The deadliest earthquake on record"

Earthquake classification—"Ring of Fire," convergent oceanic/ continental plate margin.

Magnitude—probably 8.0.

Death toll—up to 830,000.

Principal cause of death—collapse of cave dwellings in loess cliffs.

Property loss—incalculable.

The earthquake devastated a swathe of land more than 500 miles wide and affected 92 counties in the Shansi region; over 50 percent of the entire population was killed. One reason for the high death rate was the fact that a large proportion of the population lived in caves carved out of cliffs of soft loess, a wind-deposited silt-type soil that is very prone to erosion. The earthquake generated many landslides that destroyed the caves, causing them to collapse and burying the inhabitants under tons of loess.

*The classification of the volcanoes, earthquakes, and tsunamis in the panels is based on post–plate tectonic thinking, and the terms used are described in the text.

fact, the first instrument for measuring earthquakes, the seismoscope, was invented in AD 132 by Chinese scholar Chang Heng (AD 78–139). Chang Heng believed that earthquakes were caused by air moving from place to place and becoming compressed and trapped in confined spaces. In other words, he opted for a natural, rather than a supernatural, explanation.

▲

In some cultures, the myths and legends associated with volcanoes, earthquakes, and tsunamis persisted with time; but in those societies where civilization evolved, the myths began to give way gradually to more rational ideas. Nowhere was this more evident than in Western culture. But the irony here was that one straitjacket was simply replaced by another, as explanations of natural phenomena rooted in the myths of ancient superstitions were replaced by those imposed by the monotheistic Christian religion—a religion that came with its own set of constraints against scientific progress, arguing that natural disasters were either the work of the devil or a manifestation of the wrath of God. The *supernatural* had been replaced by the *divine*, and the ensuing Church vs. Science debate produced bitter conflicts, embroiling the protagonists in, at best, public ridicule and, at worst, charges of heresy.

As geologists strove to understand the evolution of the earth, they threw off the shackles of both superstition and religion. We shall see that as they did so, scientists like James Hutton, with his vision of rock cycles, slowly began to be drawn toward some kind of *great theory* that would unify and explain the various physical processes acting on the earth.

Could one theory be so all-embracing? In the years of the second half of the twentieth century, many geologists would say maybe yes. What brought about this mindset? The answer is plate tectonics—the earth science "theory of everything."

Volcanoes are a conduit through which molten material from inside the earth reaches the surface of the planet. Earthquakes are a result of the release of strain in the earth's crust. Two different kinds of phenomena. But it had been known for many years prior to the theory of plate tectonics that there are similarities in the distributions of volcanoes and earthquakes on the surface of the planet. Could it be that the *processes* behind volcanoes and earthquakes, although giving rise to different outcomes, are in some way linked? And could they be explained by plate tectonics? We know now that the answer is yes. But this is only part of the story.

Not only can volcanoes and earthquakes be explained by plate tectonics, they both play a pivotal role in the process itself.

So how did earth scientists arrive at this conclusion?

To find an answer to this question, we must travel a road that stretches from myth to state-of-the-art science. But was it a linear progression? The answer is, of course, no. Far from it. It was more of a stumble, really, and for many years geologists did not even dream of finding a theory of everything. So what kind of journey *was* it, this road to plate tectonics? For one thing, it was long and winding, and there were dead-ends aplenty. But as scientific thought developed, three monumental ideas became almost unstoppable.

The first was that *natural* processes, as opposed to divine acts of God, worked on the surface of the earth.

The second was that these processes were somehow linked together in great *cycles*.

The third was that nothing was *permanent*—that great mountain ranges had come and gone, and perhaps even the continents themselves had drifted around the surface of the planet. This was perhaps the most breathtaking concept, but it was still only one milestone, albeit it a very important one, on the road.

Now, let's turn our attention to how scientists arrived at the greatest ever breakthrough in earth sciences—the theory of plate tectonics.

2
BEGINNINGS

How is it possible that the continents, the solid earth we live on, can drift around the surface of the planet? And what great forces must be at work to drive the movements of such large blocks of material?

The answer to both these questions lies in the theory of plate tectonics—the theory that finally destroyed the myths surrounding volcanoes, earthquakes, and tsunamis. But until around the beginning of the last century, it was widely believed that the positions of the continents and the oceans were permanently fixed on the surface of the earth. Now it is known that they are constantly changing their positions. This implies a major reversal in the way we perceive the history of the planet. But it is a reversal that has been a long time in the making.

Great battles over systems of belief have been fought between Science and the Church for centuries, the prize being the way in which we view our place in heaven and on earth. In these battles religion has been challenged, at one time or another, by most systems of formal science. Galileo's universe, in which the earth orbited around the Sun, and Darwin's vision of an evolution governed by natural selection are prime examples of such challenges on a grand scale. The battles associated with resistance to ideas such as these have been at the heart of the development of many of the important philosophical as well as scientific aspects of Western thought.

In a sense, the road to plate tectonics theory originated back in the time when ancient myths were starting to be replaced by religious beliefs, and the enormity of the concept can only be truly appreciated if we view it in this historical context.

▲

Geology, the science of the earth, has had more than its fair share of battles with the church. One reason is that geology, more than most sciences, has always been concerned with change in the natural order of things—an idea that was for many years, and to some extent still is, at odds with the views of orthodox religion. The resulting Church vs. Geology conflict therefore centered on the religious establishment's objection to change in the natural world.

Any kind of change, it would seem.

But what really frightened the leading churchmen was the idea of long-term change associated with large-scale planetary evolution; change that could challenge the view that the formation of mountains or the placement of the ocean basins were anything but a permanent one-off result of creation. In other words, the church feared the recognition of change that could have reshaped the face of the earth.

There can be no more ultimate example of geological change than the movements of the continents. So to oppose, once and for all, any objection to geological change we must ask three key questions:

1. What is the structure of the earth, and does it allow the continents to move?

2. How old is the earth, and has there been time for the continents to move?

3. Has the earth actually changed during its history?

Taken together, the answers to these questions encompass many of the advances made in geology over the past several hundred years. To understand this, we must trace the way the science has moved forward. But we will not attempt to explain geology within the framework of a rigid chronology, nor give an all-encompassing history of the science. Instead, we will concentrate on those advances that have impacted, directly or indirectly, on the developments that finally led to plate tectonics theory. These advances will be explored as a "caravan of ideas"—ideas that intermixed and gained from each other, until they finally fused into the broad system of scientific knowledge that we know today. And as the individual strands of this knowledge are traced, we will encounter some of the great figures in the history of the natural sciences.

WHAT IS THE STRUCTURE OF THE EARTH?

If large-scale movements of parts of the earth's surface have taken place, it follows that the planet must have a structure that allowed this movement to happen.

Many of the original ideas on the structure of the earth, such as the Chinese belief that dragons lived below mountains, were rooted in folklore. Others were embedded in deeply held religious beliefs, such as the idea that there was a fiery hell down below peopled by wicked souls. The latter idea was challenged by Thomas Burnet (1635–1715), who proposed that there were ample stores of water, which had retreated from the Flood, lying below the surface of the earth. Some of the concepts that followed seem to me no less bizarre. For example, John Cleves Symmes (1780–1829) suggested that the interior of the planet was hollow and contained a number of habitable concentric spheres. But, bizarre or not, Symmes was right in one respect—the earth's interior does consist of a series of "layered spheres."

Our present knowledge of the structure of the interior of the earth is based on geophysical data, obtained largely from the study of the speed of earthquake waves as they travel through different kinds of material at different velocities. Essentially, in the modern view, the earth consists of three layers, with discontinuities between them.

The *crust*, or thin surface layer, has a thickness of approximately 40 kilometers in continental regions, but only around 6 kilometers under the oceans. At the base of the crust lies a boundary termed the Mohorovicic discontinuity (the Moho), which was discovered by the Yugoslavian geophysicist Andrija Mohorovicic in 1909.

The *mantle*, which makes up the bulk of the earth (75 percent to 85 percent by volume), extends from the Moho to a depth of approximately 2,900 kilometers. The mantle is composed of material with a temperature of 1,500°C to 3,000°C that contains lighter elements, such as oxygen and aluminium. The inner part of the mantle is capable of slowly *flowing* in the form of convection cells—very important, as we shall see later. The Gutenberg discontinuity lies between the lower mantle and the core.

The *core*, which has a radius of around 1,070 kilometers, is composed of a mass of hot, dense, magnetic material rich in the heavier elements, iron and nickel. The core can be subdivided by a boundary identified by Inge Lehmann, a Danish seismologist, in 1936. In the center, or inner core, the material is in a semisolid state at a temperature of approximately 5,000°C; in the outer core it is more fluid, at a temperature of approximately 4,000°C. Scientists think that the heat that keeps the core hot is provided mainly by the radioactive decay of elements such as uranium and thorium. Eventually, these elements will run out and, as they do, the earth will become a cold, lifeless planet.

This, then, is the basic three-layer structure of the earth. But, in addition, there is an important zone called the *lithosphere*. This is a solid zone that includes the crust of the earth, together with the uppermost part of the mantle. The lithosphere is considerably thicker under the continents (up to 100 kilometers) than under the oceans (usually less than 10 kilometers). At this stage, therefore, we can make a fundamental distinction between the type of lithosphere lying beneath the continents (thicker) and that lying below the oceans (thinner). The lithosphere is directly underlain by the *asthenosphere*, the upper part of the mantle that is soft and partially molten. The asthenosphere is sufficiently hot for a small amount of melting to take place. It is also under less pressure than the rest of the mantle, so that it becomes deformable or plastic and is thought to flow. As a result, the lithosphere is effectively detached from the mantle below and can move independently on the asthenosphere. In this respect, the asthenosphere is critically important as a medium of transport because as it flows it permits the movement of the lithospheric material lying on it.

It is therefore apparent that the structure of the earth is such that it does provide a mechanism whereby large-scale movements of blocks of lithospheric material, such as the continents, are possible.

At least in theory.

HOW OLD IS THE EARTH?

So, in terms of the structure of the earth, it is possible that blocks of lithosphere can move via the medium of the asthenosphere. But this is not the only requirement needed for the movement to actually occur. Clearly, the apparently changeless location of the continents and oceans over recorded history tells us that if such movements do occur, then they are slow and require time. A great deal of time.

One of the earliest and most fundamental questions posed in all societies involves the nature of creation. The earliest creation stories were deeply embedded in myth. The Greeks believed that time had no beginning and no end, whereas the Hindus favored endless cycles of destruction and renewal. In much of Western culture, the creation story was based on the biblical version described in Genesis, and it is important to understand that in this account there was a definite beginning—a single point in time at which creation started. Furthermore, in Christianity, time was *not* thought of as being cyclical with events repeating themselves endlessly.

The most famous of the Bible-based estimates for the date of the beginning of creation was that made in 1650 by James Ussher (1581–1656). Ussher was the archbishop of Armagh, and in his *Annals of the World*, he defined an incredibly detailed chronology in which creation began at 9:00 a.m. on Monday, October 26, 4004 BC, and was completed six days later when everything on the earth was in place. In this time frame, the date of the Flood was set at around 2300 BC.

Although it seems extraordinary now, looking back from the standpoint of modern science, Ussher's estimate was still in vogue until the eighteenth century—and even later in some church circles. An important consequence of setting such a rigid time frame for creation was that it imposed severe restrictions on scientific thought. One of the most important of these was that according to the Genesis account, the features of the earth were formed at the time of creation and have *not changed* since. Except, of course, as a consequence of the Flood. This concept of permanency was a rigorously held belief that dominated religious thought in the West for hundreds of years, and it played a pivotal role in the Church vs. Geology debate.

In the seventeenth and eighteenth centuries, there were a number of attempts to estimate the age of the earth using a variety of quasi-scientific approaches. At first there was a reluctance among natural scientists to dismiss the Ussher date out of hand, and one suggestion was that there might have been a series of "multiple creations" so that the 4004 BC date need only have applied to a final stage in the creation story. This was followed by several attempts to adjust Ussher's date by thinkers who, although they attempted to be innovative, still invoked the control of nature by an all-powerful Deity.

A significant advance was made when Benoit de Maillet (1656–1738), a French diplomat, derived an estimate of over 2 billion years for the age of the earth. This was based on cosmological grounds—specifically that the earth had been a star that had burned out—and Maillet was well ahead of his time in predicting an age for the earth of billions of years. Another attempt to predict an age for the planet on the grounds of cosmology was made by Comte Georges-Louis Leclerc de Buffon (1707–1788). This French naturalist proposed that the planet had formed from a white-hot mass torn from the Sun and assumed he could derive an age for the earth on the basis of how long it had taken to cool—an idea first proposed by Sir Isaac Newton. To test this view experimentally, Buffon heated two large iron spheres until they were red hot and scaled up the time they took to cool down. In the end, he derived an age of between 50,000 to 70,000 years for the earth, an estimate first published in 1749 in his *Histoire Naturelle*.

Over a hundred years later, in 1882, Belfast-born scientist Lord William Thomson Kelvin (1824–1907) also used the cooling history of the planet to derive an age for the earth of 98 million years, an estimate later revised downward to between 20 million and 40 million years. The principle was sound enough (i.e., how long had it taken the earth to cool from the initial state when it was formed?). Unfortunately, age calculations based on Kelvin's concept of the cooling of the earth were completely misleading because at the time, a major source for *heating* the earth's crust—radioactive decay—had not yet been discovered.

Other lines of evidence were emerging to support the view that the earth was very old. An example was Darwin's theory of natural selection, which demanded many millions of years for evolution to work. Toward the end of the nineteenth century it was, therefore, generally agreed among natural scientists that the earth was at least millions of years old. Yet the need for a reliable method of accurately measuring such an immense time span was as great as ever. Ironically, perhaps, that need was to be met by the heat source that was unknown in Kelvin's time—radioactive decay.

The key breakthrough in the discovery of the radioactive clock was the 1896 finding by Henri Becquerel that uranium-bearing minerals in rocks emit radioactivity—a term first suggested by Marie and Pierre Curie in 1898. Radioactivity involves the continuous emission of nuclear radiation from the nuclei of atoms, which shoot out tiny particles of matter (alpha and beta particles, and sometimes bursts of wave energy termed gamma rays) as they disintegrate—a process known as *radioactive decay*. During this process, one element (the parent) can change into another (the daughter), or even into a series of daughters. In this way, the uranium decay series ultimately produces lead, which is stable; that is, not subject to further radioactive decay. In 1921, Henry Russell made one of the first chemical estimates of the age of the earth using radioactive techniques and derived a maximum figure of 8 billion years, based on the total uranium and lead in the earth's crust. In 1927, Arthur Holmes, author of *The Age of the Earth,* used a different value for the total uranium and lead in the earth's crust and came up with an age for the earth of between 1.6 billion and 3.3 billion years.

Later, the chemical age was to be replaced by the *isotopic* age. Put simple, the nucleus of an atom contains positively charged protons and electrically neutral neutrons, and is surrounded by negatively charged electrons. The number of protons in the nucleus (the atomic number) is fixed, but most elements have a variable number of neutrons so that the atomic mass, or atomic weight, of the element varies. An *isotope* is one of two or more forms of an element that have the same atomic number and are chemically identical, but have different atomic weight and nuclear properties. Isotopes can be either stable or radioactive; the latter undergo radioactive decay and are termed *radioisotopes*. The rate at which a radioactive isotope undergoes decay is expressed in terms of a *half-life*—the time it takes for half the nucleus to change into another isotope. The principle underpinning radioactive dating is that the rates of decay of many isotopes have been very tightly determined, so that if the parent/daughter ratio can be measured and if the half-life of the isotope is known, then the time it takes to reach that radioactive state can be calculated to give the age of a sample.

The isotopes of lead in a sample can be used to calculate the age of the earth if two provisos are met. First, the sample must have been formed at the same time as the earth, and second, the isotopic ratios of primeval lead at the time the earth was formed must be known. At first, meeting both of these provisos proved difficult, until the answer was found in meteorites. Meteorites have a common origin with the earth and contain the type of lead that was present in the solar system when the planet was first formed. The American geochemist Clair Patterson derived "primeval" lead isotopic ratios from a meteorite containing only very small quantities of uranium, so that the lead isotopic ratios had not been changed by uranium decay. Using this information as a baseline, Patterson's 1956 estimate of the age of the earth, based on lead isotopic ratios, was around 4.55 billion years, a figure that is still accepted today. This is a very long way from Archbishop Ussher's estimate, made more than 350 years ago, and follows a general progression for the age of the earth from thousands to millions to billions of years. On the basis of the new dating, we can now draw up a geological timescale (see Table 2-1).

Thus, the latest "age of the earth" estimate provides an ample timescale for very long-term changes to have taken place on the surface of the planet since it was originally formed. Now we come to the most crucial question of all.

HAS THE EARTH CHANGED DURING ITS HISTORY?

It would appear that neither the physical structure of the earth, nor the time available since the planet was formed, stand in the way of large-scale change. So has such change actually occurred?

The answer is *yes*, of course it has. That seems obvious to us now. But it was not always so, and as the Science vs. Church battle was fought, it focused on the question of what had happened to the earth *after* it was formed. In this way, the conflict went far beyond simply setting a date for creation.

Essentially, two opposing views, both stemming from the age-

TABLE 2-1
A Simplified Geological Timescale

Eon	Era	Period	Epoch	Age (at start)
Phanerozoic	Cenozoic	Quaternary	Holocene	
			Pleistocene	1.6 million years ago (Ma)
		Tertiary	Pliocene	
			Miocene	
			Oligocene	
			Eocene	
			Paleocene	65 Ma
	Mesozoic	Cretaceous		140 Ma
		Jurassic		205 Ma
		Triassic		250 Ma
	Paleozoic	Permian		290 Ma
		Carboniferous		355 Ma
Devonian				410 Ma
Silurian				438 Ma
		Ordovician		510 Ma
		Cambrian		540 Ma
Precambrian				4,550 Ma

of-the-earth debate, came into collision. On the one side were the "creationists," followers of Ussher, with their deeply held view that the earth was young and had been created intact as described in Genesis and, equally important, that all the earth's features had been fixed at that time and have remained fixed ever since. The only exception were the changes that followed Noah's Flood. Indeed, the Flood was often used by the church, together with the accusation that some things were the work of the devil, to explain phenomena—such as the occurrence of fossils—that would otherwise have proved to be difficult, if not impossible, to rationalize.

Two spectacular events could therefore be said to have dominated the thinking of the creationists: *Creation* itself and the *Flood*—one that gave birth to the earth, and the other that changed it dramatically in a "once and for all" manner.

Lined up against the creationists were the "natural scientists"—inheritors of the thinking of Rene Descartes and later Sir Isaac Newton, both of whom in their own ways had looked for secure foundations for knowledge in the belief that everything could be explained by physical causes. In a kind of post-Newtonian desire to find order in nature, the natural scientists subscribed to the view that the earth was vastly older than had been predicted by Ussher and, furthermore, that it was in a state of *constant* change.

One of the main barriers that the natural scientists had to overcome was the idea that everything in the history of the earth had been a direct result of divine intervention, and on these grounds especially the battle between the creationists and the natural scientists was fiercely fought. It brought the scientists into conflict not just with the church, but also with many other parts of an establishment that believed in the permanency of the social, as well as the natural, order of things—a layering of society in which every person had a fixed state "under heaven and earth."

For me, in the twenty-first century, it is difficult to get inside the mind of one of the early natural scientists. For one thing, it is hard to understand that religious constraints could impose such a straitjacket on the way the natural scientists were allowed to think. For another, it is difficult to comprehend how much they feared the consequences of heresy, a fear that was real enough, as it turned out. But the real stumbling block to identifying with the natural scientists was that many of them did not *want* to dispute biblical beliefs. And as a result, a number of ingenious attempts to sidestep the constraints emerged.

Nicolaus Steno (1648–1686), a Danish physician and natural scientist, was one of the earliest geological thinkers. Steno, who was originally called Niels Stensen, studied fossils and made a number of radical observations about the crust of the earth. One of these was that horizontal layers, or strata, can be found lying one on top of top of the other, and that the first in a sequence should be the oldest. Furthermore, the different strata often had a different population of fossils. Steno also recognized that in some places the strata had been distorted, which, he believed, was the result of changes brought about by forces such as rifting. He also speculated that the forces involved in rifting could raise and lower mountains. In this respect, Steno was advocating massive changes to the earth's surface. In 1669, he published his work *Prodomus,* in which he attempted to decipher the history of the earth's surface on the basis that the past could be reconstructed in terms of what was happening at the present. This was a landmark because it was one of the first expressions of what was to become one of the cornerstones of geology—that *the present is the key to the past.* Because of his work, Steno was later called the founder of geology, yet for many years he was largely ignored.

Another seminal work in early geological literature was *The Sacred Theory of the Earth,* published in 1681.The author was Thomas Burnet, an Anglican clergyman and royal chaplain to King William III of England. Burnet followed the teaching of Ussher and believed that the earth was created some time around 4000 BC, with the Flood occurring 1,600 years later. The Flood occupied a central role in Burnet's history of the earth, and according to him, the features of the planet's surface were formed at the time of the deluge. But Burnet encountered a problem in finding enough water to submerge the land during the Flood. In the end, he overcame this issue by suggesting the Flood resulted from the surface of the planet cracking open and releasing waters that were present as a layer in the interior, a scenario consistent with the Scriptures.

According to the model proposed by Burnet, the antediluvian earth was a smooth ovoid, or egg, with a uniform texture. When the Flood came, sections of a weakened crust cracked open, exposing the water layer beneath. Some of the cracked crust fell into the interior and the floodwaters burst onto the surface. In this process, mountains were upturned fragments of the early crust left behind after the deluge, and the oceans were holes left where the crust broke up. This was one of the first theories to attempt to explain how the principal features of the earth's surface were formed on a global scale.

Far-fetched by our modern scientific standards? Certainly. But in several ways Burnet was ahead of his time. He suggested that particles could somehow settle out of an ocean to form rocks. Furthermore, he refused to rely solely on divine intervention as the force behind the formation of the planet. Instead, he proposed that the earth had been formed slowly by *natural* causes, and that alterations to the surface were the result of nature's agents such as running water, winds, and frosts—another extremely important concept underpinning geology.

Although he questioned the degree of divine intervention in shaping the earth, Burnet did try to reconcile science and religion. And almost inevitably, given the then-current climate of thought, he failed. He believed that nature was God's work, and that one truth (God's) cannot be in opposition to another (nature's). But in the end, he sidestepped the issue by invoking a get-out clause that states that discrepancies between religion and science would be reconciled by future discoveries. However, like so many of the early natural scientists, Burnet suffered for his beliefs, and by challenging part of the biblical doctrine he had to forfeit any ambition he had for advancement in the church.

Another significant contribution to early geology was made by Buffon, the same man who heated the iron spheres to assess the age of the earth. He divided the history of the planet into seven epochs ranging from the formation of the solar system to the appearance of man. He also provided a remarkably advanced insight into a kind of biological evolution, claiming that all animals had descended from a single living being. Unlike Burnet, who tried to work the scriptures into his theory, albeit by stretching them too far in the opinion of his peers, Buffon rejected the Bible. Instead, he applied an evolutionary approach to the formation of the earth.

The challenge to the restrictions of the biblical fundamentalists was truly out in the open now.

▲

During these early years in the development of geology, theories abounded. The shackles imposed by the creationists had been severely challenged, and the first attempts at a global-scale rationale to explain the features of the surface of the earth had emerged. But any understanding of the processes behind the theories was still fanciful, and empirical evidence (based on observation or experiment) was sparse. What was needed for the infant science of geology to advance was the foundation of a systematic body of factual knowledge.

A number of great innovative thinkers moved geology toward this end, including James Hutton, Abraham Werner, Georges Cuvier, William Smith, Charles Lyell, and Louis Agassiz—a list of some of the giants in the natural sciences. But before moving on to assess the work of these scientists, we need to understand something of the nature of the rocks that form the earth's crust.

3
NEW BATTLES

THE ROCKS OF THE EARTH

Many complex schemes exist for the classification of rocks, but for us it will be sufficient to simply distinguish between three major types of rock in the lithosphere: igneous, sedimentary, and metamorphic.

Igneous Rocks. Named from the Latin word for "fire," they are primary rocks formed of minerals derived from cooling magma, the molten material within the earth. Two of the most widely used parameters for the classification of igneous rocks are those based on 1) their chemistry and mineralogy, and 2) their method of formation, or emplacement.

At this stage, the classification of igneous rocks on the basis of their chemistry and mineralogy can be greatly simplified by introducing the terms SIAL and SIMA, which will be used later to describe the nature of the crust underlying the *continents* (light, granite-rich SIAL) and *oceans* (heavy, basalt-rich SIMA).

SIAL refers to high-silica, light-colored igneous rock that's rich in the elements s̲ilicon and a̲luminium and in minerals such as silica (SiO_2) and alkali feldspar (Na, K, and Al silicate). An example of this kind of rock is granite, an important constituent of the continental crust. SIMA refers to low-silica, dark-colored, igneous rock that's rich in the elements s̲ilicon and m̲agnesium, which are contained in minerals such as pyroxenes (Mg, Ca, and Fe silicate), olivine (Mg and Fe silicate), and plagioclase feldspar (Na, Ca, and Al silicate). These rocks include basalt, which is the major constituent of the oceanic crust, and gabbro.

There are two principal ways in which igneous rocks can be emplaced:

- ▸ Intrusive igneous rocks are formed below the earth's surface either in magma chambers or following intrusion between overlying strata. The deepest intrusive igneous rocks are called plutonic, and they are composed of large crystals that have had time to cool slowly. Examples of these rocks include granite and gabbro.

- ▸ Extrusive, or volcanic, igneous rocks are extruded at the earth's surface, either into the air or underwater, where they cool quickly—a process that results in the rocks having very small crystals and a glassy appearance. The major example of a volcanic rock we will come across is basalt.

Sedimentary Rocks. These are composed of material that has been laid down by water, wind, ice, or gravity. Sediments can be divided into three groups:

1. *Clastic sediments* are made up of fragments of preexisting material usually weathered from other rocks. Clastic sediments include sandstone and clay.

2. *Chemical precipitates* are formed of material precipitated from solution, and include some limestones and evaporates.

3. *Organic sediments* are composed either of the fossilized remains of organisms (e.g., as found in many limestones) or organic matter (e.g., peat).

   If the material forming the sediment has become consolidated, the deposit is referred to as a sedimentary rock.

Metamorphic Rocks. They are formed from the action of heat and/or pressure on preexisting rocks, a process that can substantially alter both the mineralogy and the structure of the original rock. Often particular suites of minerals are associated with particular temperature and pressure conditions, so metamorphic rocks can offer an indication of the pressure and temperature conditions under which they were formed.

THE BATTLE IS JOINED

We have seen that as geology developed, three ideas gradually began to gain ground.

The first was that the earth was much *older* than had previously been claimed.

The second was that the earth had undergone immense *changes* since it had been formed.

The third was that the changes that had affected the earth were entirely *natural* and had not resulted from divine intervention.

The importance of these three ideas to the progression of the science of geology cannot be overestimated. In particular, it meant that as geologists attempted to move the science forward they were no longer confined within the straitjacket of a Genesis-type creation that set unacceptable time limits and fixed forever the features of the earth.

The Church vs. Science debate had one of its fiercest battlegrounds in the Lisbon earthquake and tsunami of 1755 (Panel 5). The earthquake and subsequent tsunami took place on a Catholic holiday and many of the churches that were destroyed were filled with worshipers. The religious leaders of the day believed the disaster came from the anger of God—that it was a divine punishment. But there were attempts to find a more rational explanation. For example, the German philosopher Immanuel Kant (1724–1804) sought a scientific rationale for the event and suggested that the earthquake resulted from movements in subterranean caverns full of hot gases that were part of the earth's respiratory system. This wasn't true, of course, but the important point is that Kant offered an explanation based on a natural phenomenon. In addition, the prime minister at the time, Sebastião de Melo, collected data on the time of the disturbance and damage caused from the priests in the surrounding areas. In some ways, their efforts were a forerunner of modern seismological practices.

PANEL 5
THE LISBON EARTHQUAKE AND TSUNAMI*
(PORTUGAL, 1755)

"The largest and most destructive documented tsunami in the Atlantic Ocean"

Tsunami classification—earthquake generated.

Location of earthquake—North Atlantic Ocean.

Earthquake magnitude—estimated 6.8 or greater, possibly even up to 9.

Tectonic setting—disputed. The earthquake possibly occurred along the Azores-Gibraltar Fracture Zone between the African and European tectonic plates, with a possible zone of subduction below Gibraltar.

Overall death toll—approximately 60,000, but may have been as high as 100,000.

Principal cause of death—direct earthquake damage from collapsed buildings, fire, and the effects of the tsunami.

▲

In the eighteenth century Lisbon, the capital of Portugal, was one of the most beautiful cities in Europe, with a population of around 250,000. The earthquake of 1755 struck at around nine-thirty on the morning of the first of November—All Saints Day. The city was totally unprepared.

The exact location of the earthquake epicenter is not known with certainty, but one estimate puts it in the Atlantic off the southwest coast of Portugal, around 200 kilometers to the southeast of the city. There were three shocks, with the second being the strongest, and for nine minutes after it the ground continued to shake, structures swayed, and fissures opened up in the ground. Buildings, weakened now by the ground swaying, began to collapse, and more than 80 percent of those structures in the city were destroyed. It was a religious festival, with many people at worship in the churches, and Lisbon's five great cathedrals collapsed, killing thousands.

The shaking stopped, and the inhabitants of the city began to think the worst was over. Until about thirty minutes after the earthquake struck, a tsunami approached the coast and began to run up the Tagus estuary. As it reached the shallow waters it grew in height, and estimates put one of the waves at more than 30 meters as it moved along the estuary toward Lisbon. Ironically, many people had fled to the harbor, where some had boarded ships, and others crowded the banks of the river thinking to escape falling debris and fires—only to be hit by the wall of water from the tsunami.

When it struck, the tsunami caused considerable confusion because it came in a series of three waves, not a single one. To add to the chaos, at the start of the event the water was dragged back far enough to reveal old shipwrecks in the harbor. Then the tsunami came in along the river and crashed into the harbor, swamping ships moored there. Beyond the harbor, the water engulfed the lower part of the city on either side of the river, and at the end of the tsunami, many buildings, including the royal palace, were destroyed. After the tsunami came the fire, which ravaged Lisbon for five days, destroying much of what was left of the city. The combination of earthquake, tsunami, and fire was a pungent cocktail.

Within a year, Lisbon was reborn, a modern capital to replace the medieval city destroyed in the events of 1755. One disturbing consequence of the earthquake was that because it had struck a Catholic country, some people began to question their faith, and this at a time when religious and rational systems of belief were in severe conflict.

The Lisbon tsunami was transmitted through the North and South Atlantic and out into the Indian Ocean and the

Pacific. The strongest effects were felt in the North and South Atlantic, and areas that suffered the worst loss of life and damage to property from the tsunami included the Portuguese and Spanish coasts, Gibraltar, Tangier and the western coast of Morocco, and Madeira and the Azores in the North Atlantic.

*The classification of the volcanoes, earthquakes, and tsunamis in the panels is based on post–plate tectonic thinking, and the terms used are described in the text.

In the Lisbon earthquake of 1755, the data collected was crude, and it took another hundred years for the next major breakthrough in earthquake studies to emerge. This happened in 1855, when the Italian scientist Luigi Palmieri invented his seismograph, a device that picked up ground vibrations in earthquakes using mercury contained in tubes. It was now possible to measure earthquakes in a more accurate manner.

The Lisbon event was thus a watershed in the study of earthquakes, and although the religious view continued to be supported in many circles, the gathering of hard data meant that things would never be quite the same again.

Inevitably, however, even as the shackles were thrown off, new conflicts arose as geologists argued among themselves about how the natural changes to the face of the earth had occurred. At the onset, two schools of thought began to take shape—one led by the *uniformitarianists* and the other by the *catastrophists*.

Uniformitarianism emerged largely from the work of James Hutton (1726–1797). Although Leonardo da Vinci, some 300 years earlier, also made similar observations, it is Hutton who is regarded by many as the real founder of what was then "modern" geology. In essence, uniformitarianism may be defined as the theory of *gradual change*. According to Hutton, the slow processes that are at work on the earth's surface had operated in the past and would continue to operate in the future; a suggestion of a uniformity of planetary evolution.

Hutton's ideas were published in his *Theory of the Earth* in 1795.

He believed that the planet was of great antiquity. He also believed that erosion wore down mountains and continents, and that soils were products of this erosion. So far, so good. But in my view, it was at this point that Hutton revealed his true genius because he realized that erosion, left unchecked, would simply wear down the surface of the earth, making it flat and featureless. Since this quite clearly had not happened, Hutton proposed that some counter-process must be in operation, and he suggested that the diversity of the earth's crust was maintained by a *rock cycle*.

In this cycle, rocks are eroded and reduced to particles by natural agencies, such as water and wind, and then transported around the surface of the earth. At this stage, however, the process begins to repair itself and the particles, mixed with the remains of life-forms, are deposited to form sediments that are buried in water. The sediments then undergo hardening and consolidation by processes such as baking by heat and compacting by pressure, and eventually they are uplifted to become mountains, which, in turn, are subjected to erosion and the cycle starts over again. In this radical view of the earth, mountains were not *permanent* features but would come and go with time as part of a very slow, constant re-arrangement of the features of the planet.

The idea that a "rock cycle" shaped the earth's surface was a far-reaching concept. But, as was to happen with the theory of continental drift many years later, the problem was to find a mechanism that could explain the large-scale movements involved in the cycle; in this case, the uplift of mountains. Hutton believed energy sources must reside in the earth's interior, and he concluded that internal heat was the force behind the uplift. In this way, Hutton had devised a crude early version of the continental rock cycle that is such a central part of plate tectonics theory.

In Hutton's framework, the history of the earth's surface is governed by the long, slow cyclic trinity of *erosion-consolidation-uplift*.

This was an early indication of a "dynamic earth" and was critically important to the further development of geology because it showed that at last scientists were beginning to feel their way toward the holy grail—a system in which the planet behaved as a single entity and was governed by a universal series of natural principles. The holy grail was still far off in the distance, but it had made its first shadowy appearance.

Catastrophists also agreed that the earth's surface had suffered severe alteration. They differed fundamentally from the uniformitarians, however, in proposing that these alterations were not the result of gradual change, but were brought about by sudden major catastrophes, such as floods, earthquakes, large-scale volcanic activity, and massive tsunamis.

The two theories could therefore be defined as one of *gradual change* resulting from evolution (a view held by the uniformitarianists) versus one of *sudden change* brought about by disasters (a view supported by the catastrophists).

The battle between the two schools of thought took another turn toward the end of the eighteenth century when a second major debate, of which Hutton was also a part, raged among geologists. This time the main protagonists were the *Vulcanists* (or Plutonists), centered in Scotland and driven forward by Hutton, and the *Neptunists* located in Germany and led by Abraham Werner (1750–1817), professor of mineralogy at the University of Freiberg. Much is apparent from the naming of these two theories: the Neptunists following Neptune, the Roman god of the sea, and the Vulcanists following Vulcan, the god of fire.

For the Neptunists, Werner arranged the crust of the earth into formations. He then distinguished five major formations that he proposed corresponded to different ages in the history of the earth. Each formation was subdivided into a number of layers, or strata. Following Nicolaus Steno, Werner made a link between age and the sequence in which the strata had been deposited, and his system of the classification of strata became the model for all future stratigraphic systems in geology.

Werner could be considered an important pioneer in geological thinking, and his 1787 publication, *Short Classification and Description of the Rocks*, is a seminal work. However, a problem arose when Werner attempted to explain the origin of his formations. He proposed that at some time in the past, the core of the earth had been covered with a primeval ocean out of which precipitated, layer by gradual layer, all the rocks forming the present-day crust. In this concept, therefore, water (Neptune) was the cause of all geological formations, including rocks such as granite.

In contrast, Hutton, while still accepting the role played by water in the evolution of the earth, put considerable additional emphasis on the influence of the red-hot interior of the planet, symbolized by the fire god Vulcan. According to the Vulcanists, rocks like granite were igneous; that is, they had not precipitated from water, but rather had crystallized from hot molten magma. Neptune vs. Vulcan, or water vs. fire, was a fundamental argument in the history of geology.

Werner was supported by fellow Neptunist Richard Kirwan (1773–1812), a founder and fellow of the Royal Irish Academy. Kirwan was a great believer in the Flood as a geological agent, and he strongly disputed the idea that granite was formed from a hot magma. He challenged Hutton on both scientific and religious grounds—even to the extent of accusing him of heresy, which was a very serious charge to level against a pious Quaker like Hutton.

Another opponent of Hutton's was the meteorologist Jean Andre Deluc (1727–1817), a catastrophist who opposed the view that present geological processes had acted in the past. Deluc still retained a strong biblical element in his geology, but was prepared to compromise by going along with the idea that the six days of creation were, in fact, six epochs of much longer duration. His ideas were marshaled in his book, *Trail Elementaire de Geologie*, which, when published in 1809, was a severe attack on Hutton and the Vulcanists.

Despite other differences, both the Neptunists (Werner) and the Vulcanists (Hutton) accepted that the earth had evolved slowly over a long period of time, and the debate on Ussher's creation date had finally drawn to a close—at least among scientists.

NEW HORIZONS

As the study of rock strata progressed, one of the ways in which the science moved forward related to the way fossils were viewed. One of the most contentious arguments in the Church vs. Geology conflict had for many years centered around fossils, and indeed philosophers had argued over the origins of these puzzling phenomena since at least the time of the Greeks.

Fossils are the remains of once-living organisms, and they range from the gigantic skeletons of dinosaurs to traces of the most primitive life-forms. Fossils incited religious debate because geologists eventually claimed that they provided evidence of an evolutionary scale in the history of life. This argument was often refuted by the creationists who claimed that fossils were put in rocks by the devil to confuse believers. For those fossils that were found in strata on mountaintops, it was further suggested that they had been washed there after being destroyed during the Flood.

Once scientists had accepted that fossils were, in fact, the remains of organisms, important implications surfaced. However, neither Hutton nor Werner paid a great deal of attention to fossils, and it was left to William Smith (1769–1839), an English drainage engineer, to take a great step forward when he realized that since different strata contained different communities of fossils, they could be used to yield relative dates for the ages of sedimentary formations. The culmination of Smith's work was the production of the first geological maps of England and Wales.

Georges Cuvier (1768–1832) was born forty years after Hutton and became one of the standard-bearers of the later catastrophists. Cuvier did not subscribe to the idea that one geological age passed gradually into the next, and in 1812 he published *Essay on the Theory of the Earth*, in which he used evidence from fossils to support the idea that over time there had been at least four catastrophic floods, following each of which God created new forms of life.

This upsurge of catastrophist thinking was opposed by one of the all-time great uniformitarianists, Charles Lyell (1797–1875), the man who, more than any other, inherited the mantle of Hutton himself and probably did as much as anyone to finally win the debate with the catastrophists. Like Hutton, Lyell proposed that the geological features of the earth could only have been formed very, very slowly by gradual and still-operating processes such as erosion and sedimentation. But he went further than Hutton when, in his *Principles of Geology*, published in 1830, he spelled out the principles of uniformity. One of these was that the earth had been much the same in the past as it is at the present. Another was that the processes that acted on the present earth were the same as they had been in the past, and that they had always operated at the same rates. Lyell's philosophy was therefore another example of the idea, attributed many years later to Archibald Geikie, that "the present is the key to the past."

In the end, though, the framework laid down by Lyell, which more or less rejected drastic change out of hand, involved, as it turned out, too many progress-retarding straitjackets that had to be confronted before geology could really start to come of age.

An interesting offshoot of the "gradual change" theory came in 1837, in a lecture given to the Swiss Society of Natural History by Louis Agassiz (1807–1873), who introduced the concept of the Ice Age. This was important because it described a world in which great ice sheets had covered the land from the North Pole to the Mediterranean Sea. Ice was recognized as being as important as water in shaping the surface of the earth, but in a much faster manner that perhaps overrode "gradual change."

An important stage had now been reached in the journey toward the theory of plate tectonics. The creationists, the uniformitarians, the catastrophists, the Neptunists, and the Vulcanists had largely fought out their battles. The restrictions on progressive thought imposed by the short, rigid time frame of the biblical story of creation and the necessity for divine intervention had been removed. Geologists accepted that the earth was very old and that it had undergone immense changes brought about by natural events. Some of the changes had undoubtedly resulted from rapid catastrophic events, but most had been carved out by gradual and long-lasting processes.

And so geology was moving on as the protagonists fought to impose their mega theories on the science. But it wasn't all "big ideas," and the study of earthquakes was progressing on its own. It had become apparent that the acquisition of hard data was critical to the study of earthquakes, and an important step forward was taken when Elie Bertrand (1712–1790) cataloged the earthquakes in Switzerland that had occurred over the period AD 5643 to AD 1756. But although the trend at the time was definitely away from supernatural explanations for earthquakes, a number of strange ideas were still being thrown about.

Several of these ideas were based on Benjamin Franklin's experiments with lightning, and scientists such as Giovanni Beccaria (1716–1781) and William Stukeley (1687–1765) believed that earthquakes were caused by lightning storms occurring underground. Another theory was proposed by the English physicist John Mitchell (1724–1793): He proposed that earthquakes were caused by vapors from the interior of the earth that created shifting masses of rock below the surface and produced rippling movements in rock strata.

These rippling movements, or waves, are characteristic of earthquakes, and Mitchell is one of a number of scientists who have been termed "the father" of seismology because of his suggestion that the epicenter of an earthquake (i.e., the point on the ground directly above the focal point) can be located by measuring the travel time of the earthquake waves through the crust (see Chapter 11).

Despite the scientific advances made, in some societies the myths surrounding the trinity of natural disasters stayed frozen in time. One of the North American earthquake myths, which persisted into the nineteenth century, relates to the New Madrid earthquakes of 1811–1812, the largest earthquakes, outside Alaska, ever to hit the U.S. mainland (see Panel 6). In the myth, a handsome young Chickasaw chief falls in love with a beautiful Choctaw princess. But the chief, who was called Reelfoot, suffered from a twisted foot, and he was refused the hand of the princess. When the chief kidnapped the girl, the Great Spirit became angry and stamped his foot. This caused the Mississippi River to overflow and form a lake—Reelfoot Lake, in northwestern Tennessee, which was born out of the New Madrid earthquake of 1812.

PANEL 6
NEW MADRID EARTHQUAKES*
(NEW MADRID, LOUISIANA TERRITORY, MISSOURI, 1811–1812)

"The largest earthquakes outside Alaska to hit the United States"

Earthquake classification—intraplate.

Magnitude—between 7.0 and 8.0 for the four earthquakes.

Death toll—unknown, but relatively low.

Property loss—severe, but not extensive in what was a sparsely populated area.

▲

The New Madrid Seismic Zone (or Reelfoot Rift), which lies in the U.S. Midwest, consists of a series of ancient faults stretching for about 250 kilometers through five states (Missouri, Illinois, Arkansas, Kentucky, and Tennessee). From time to time the faults have been reactivated, the activity taking place relatively deep, up to 25 kilometers below the ground.

The New Madrid Seismic Zone was responsible for the largest earthquakes to hit America, outside of Alaska. Four earthquakes struck during the activity: two in December 1811, and one in January 1812, with the largest occurring in February 1812. The last earthquake had an epicenter in the vicinity of New Madrid and the place was destroyed. Other properties were damaged and the shock was felt with considerable intensity over an area of some 80,000 square kilometers.

The earthquakes damaged forests, and one significant effect was that they changed the course of the Mississippi River; for a time it had even run backward. In addition, the disturbances caused fissures to open up, landslides to occur, and the ground to warp over an area of 500 square kilometers. Overall, these great earthquakes had a very significant effect on topography, perhaps more than any other earthquake ever to strike the United States. At the time of the four earthquakes, however, the area was only sparsely populated and there were relatively few human casualties.

Today, the story would be very different because this region is now a heavily built-up area with millions living in large centers of population. And the area is still seismically active—it has been estimated that there is a 90 percent probability that an earthquake with a magnitude of 7, or greater, will occur within the next fifty years. If such an earthquake does strike, it will probably be the biggest natural disaster in the history of the United States. It would hit heavily populated conurbations such as Memphis and break the levee systems on parts of the Mississippi, leading to extensive flooding.

*The classification of the volcanoes, earthquakes, and tsunamis in the panels is based on post–plate tectonic thinking, and the terms used are described in the text.

▲

Returning to the mainstream advance of geology, it had become obvious that change was everywhere, and as the forces that shape the earth became better understood, features like mountains were no longer thought of as being permanent fixtures in the landscape. But it seems to me that the most important advance of all was an idea that had begun to gradually emerge from the background mist—namely, that the processes acting on the earth's crust were cyclic and could be codified into Hutton's great trinity of erosion-consolidation-uplift. Essentially, the breakdown of the crust, and its subsequent reconstruction, were part of a great rock cycle.

4
TOWARD CONTINENTAL DRIFT

IMPORTANT IDEAS

In the second half of the nineteenth century, scientific speculation was rife as people, especially the Victorians, tried to order nature in the same way that they ordered their own lives. A whole raft of ideas emerged that can be considered to be major milestones along the road that was to lead to the theory of plate tectonics. Some of these ideas led only to dead-ends, whereas others were ridiculed, or even totally ignored, by the establishment. But this is how science progresses, and the jockeying for position among earth scientists, with their personal prejudices and vendettas, offers a fascinating insight into the evolution of ideas.

THE CONTRACTING EARTH

The contracting earth theory offered an explanation for the way in which the surface of the planet was shaped. It was mainly the work of James Dwight Dana (1813–1895) who, in 1846, proposed that the earth was cooling and contracting, and that its "skin," or crust, responded by shrinking. Dana believed that this shrinking was responsible for many of the physical features on the earth's surface, including huge mountain ranges, which, he suggested, were no more than wrinkles in the contracting "skin."

The formation of mountains had been the subject of geological speculation for a long time. Mountain building, during which long chains of crust have been folded, overthrust, compressed, and uplifted on a massive scale, is referred to as *orogenic activity.* Orogeny has been active nine times during the history of the earth, with periods of quiet between the episodes.

James Hall (1811–1898) recognized that there was a thickening of sedimentary units, now folded and distorted, in mountains. It appeared that a preformation stage in the development of mountains was an accumulation of sediments that subsided into a linear trough as they continued to accumulate. The term *geosyncline* was first used by Dana to describe this kind of trough; a geosyncline being a thick pile of sedimentary rock deposited in a subsiding, or sagging, marine basin in the surface of the earth. The sediments are then compressed, deformed, and subsequently uplifted into a mountain chain. In this context, mountain belts were therefore deformed geosynclines, and according to Dana's theory the forces causing the folding and deformation were the result of the contracting earth. One problem with this theory was that if mountains were the result of the cooling and contracting of the earth's surface, then they should all have been formed at the same stage in the cooling sequence and all be of the same age, which was clearly not true. Despite its weaknesses, the theory of the geosynclinal origin of mountains remained in vogue in various forms for many years, until it was eventually replaced by plate tectonics.

Dana believed that the ocean basins and the continents were very different geologically, and that they were separated from each other in a permanent configuration. The first proposition turned out not only to be true, but also to be one of the foundations on which the theory of *isostasy* was based—isostasy being important later in providing a theoretical basis for the concept of continental drift. Unfortunately, Dana got it completely wrong with the second proposition, and it eventually became apparent that the positions of the ocean basins and continents were anything but fixed permanently. But despite the fact that the "contraction theory" had many flaws, it nonetheless remained a major force in geological thought until the theory of continental drift came along.

JOINED-UP CONTINENTS

A number of lines of evidence suggested that the continents had once been joined. One of these was the similarity between some species of fossils on different sides of the major ocean basins—species that could not have crossed vast stretches of water. There was, therefore, a need to explain the fact that a number of fossil species that are now confined to specific parts of the globe had once flourished on other continents—identical life-forms on continents that are now separate landmasses.

One obvious explanation, to which we shall return later, was that in the geological past the individual continents had once been part of a single landmass that had subsequently broken up. But this idea was firmly rejected by many geologists who proposed other explanations.

One of the most popular ideas was that at one time, parts of the continents had been joined by "land bridges." These connections between landmasses were composed of granitic rock, which over time sank into the basaltic crust below the ocean. The naturalist Edward Forbes (1815–1854) was a strong supporter of land bridges to explain global faunal distributions. So was Eduard Suess (1831–1914), who attempted to explain the distribution of the fossil plant *Glossopteris* throughout India, South America, southern Africa, Australia, and Antarctica by suggesting that a land bridge once joined all the continents together.

The idea that land bridges could offer an explanation for the distribution of fossils did not, however, meet with universal approval. For example, in 1858, the French scientist Antonio Snider-Pellegrini published his book *La Création et ses Mystères Dévoilés* ("Creation and Its Mysteries Unveiled"), in which he proposed that because fossil plants in the coal beds of Europe and North America were identical, the two continents had been joined together in a "supercontinent" that had broken up, perhaps as a result of the Great Flood. Snider-Pellegrini drew his now-famous diagram showing the positions of Europe and North America before (*avant*) and after (*après*) the separation. In preparing this diagram, Snider-Pellegrini was using essentially the same approach as the early mapmakers to find a "fit" between the shapes of the two continents (see Figure 4-1). Then in 1882, Élisée Reclus (1830–1905) proposed that the continents had moved by drifting, but could offer no explanation for the mechanism behind this movement. However, in my view, Reclus can be regarded as a visionary because of his important suggestion that collisions between drift-

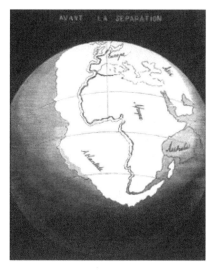
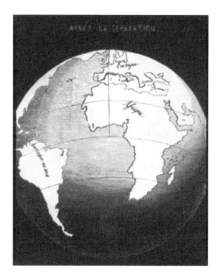

"Avant"   "Apres"

Figure 4-1. An early version of continental drift. In 1858, the French scientist Antonio Snider-Pellegrini published *La Création et ses Mystères Dévoilés* ("Creation and Its Mysteries Unveiled"), in which he proposed that Europe and North America had once been joined together in a "supercontinent" that had subsequently broken up. His now-famous diagram shows the positions of Europe and North America before (*avant*) and after (*apres*) the separation.

ing continents produced mountain ranges and left behind spaces that were later filled by oceans.

One avenue of research that supported the existence, and subsequent breakup, of supercontinents came from paleographic maps. These maps, which are based on geological (e.g., rock distribution) and biological (e.g., fossil distribution) parameters to reconstruct conditions in the past, clearly showed the existence of ancient supercontinents that have since broken up, with their fragments transported around the surface of the earth.

What, then, was the configuration of the landmasses before they began to fragment and eventually take up the positions of the present continents?

It would appear that since the earth was formed about 4.5 billion years ago, the breakup of the supercontinents has been a cyclic phenomenon. Although the first identifiable supercontinent, Rodinia, was formed around 1.1 billion years ago and may have been the supercontinent from which all others were subsequently derived, it is the supercontinent Pangaea ("all earth"), first identified by Alfred Wegener, from which the histories of the present continents can be traced. It is generally agreed that around 200 million years ago, the land was concentrated into this one supercontinent, a C-shaped landmass that spread across the equator almost from pole to pole (see Figure 4-2). Pangaea began to break up about 225 million years ago.

THE LAND BENEATH THE SEA

The science of geology was moving forward, but there was one whole section of the earth's surface that was still virtually unknown—the 72 percent of the planet covered by water. In particular, little was understood of the world lying at the bottom of the ocean. Until the HMS *Challenger* expedition changed our perception of the deep-sea.

By the end of the seventeenth century, when the heroic age of sea exploration was in decline, the *surface* of the oceans and their boundaries were largely known to explorers. However, as late as the middle years of the nineteenth century, the pre-*Challenger* time, the deep-sea was still considered by natural scientists to be a lifeless, invariantly cold, featureless abyss floored by deposits of chalk. But then new data began to slowly emerge, although in an unsystematic manner, which changed this picture.

One important survey was coordinated by Lieutenant Matthew Maury (1806–1873) of the U.S. Navy, who persuaded shipmasters to send him data on parameters such as latitude, longitude, winds, and currents from their logs, which he then entered onto charts. Many of Maury's findings were described in *The Physical Geography of the Sea*, which, when it was published in 1855, challenged old ideas and opened up new visions of the oceans. In particular, the idea of a flat featureless deep-sea was challenged when Murray produced his seabed chart of the North Atlantic in 1854. This chart hinted at the existence of three important features of the North Atlantic seabed:

1. Sea bottom depths increased rapidly beyond the edge of the continental shelf.

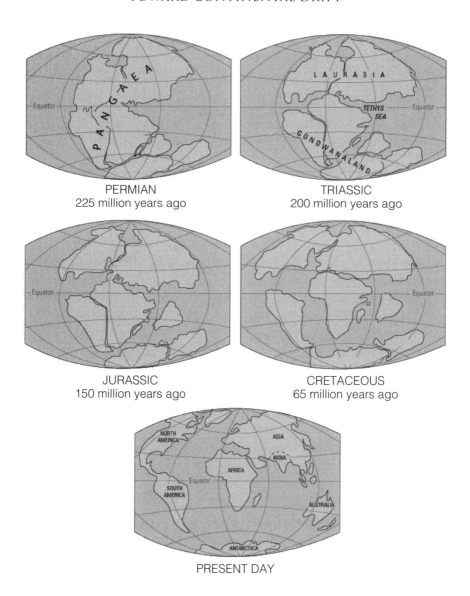

PERMIAN
225 million years ago

TRIASSIC
200 million years ago

JURASSIC
150 million years ago

CRETACEOUS
65 million years ago

PRESENT DAY

Figure 4-2. Pangaea ("all earth"). According to the theory of continental drift, Pangaea is the supercontinent that began to fragment 225 million years ago to eventually form the continents in their present configurations. Credit: United States Geological Survey.

2. Although it was not possible to define the degree to which it could be described as rugged, the seabed was not flat and featureless.

3. There was an area of elevated topography in the central part of the ocean—the first suggestion, supported by data, of a submarine mountain range in the Atlantic.

As the nineteenth century progressed, information on the deep-sea was still extremely sparse and it was becoming increasingly apparent that an integrated survey of the oceans was a scientific necessity. Oceanography is the "science of the seas," and if there is a single event that can be said to have given rise to its birth, it was the *Challenger* expedition.

This British voyage of deep-sea exploration, which was the result of the cooperation between the British Admiralty and the Royal Society of London, was conceived by Professor Charles Wyville Thomson (1830–1882). The expedition began its circumnavigation of the World Ocean in December 1872 and returned home in May 1876. The prime legacy of this momentous expedition was to integrate the chemistry, biology, geology, and physics of the oceans into the unified science of oceanography.

From our point of view, the knowledge acquired about the environment at the bottom of the deep-sea is the most important legacy of the *Challenger* expedition. Two discoveries, in particular, were of special importance.

First, Lieutenant Maury had previously shown that the seabed was not a flat world devoid of features and had hinted at the existence of an elevated area in the central Atlantic. The presence of this elevated region, where the seabed was raised up and water depths were less than 2,000 fathoms (3,660 meters), was confirmed by the *Challenger* expedition. Data obtained from the ship also suggested that such elevated areas may be present in all the major oceans. This was the first hint that a *mid-ocean mountain range* might exist on a global scale, rather than just in the Atlantic.

Second, data from the *Challenger* expedition showed that in addition to the elevated areas, or highs, the seafloor was cut by a number of lows. At these deep-sea trenches, the seabed dropped to great depths. For example, in the Challenger Deep, which is the deepest part of the Marianas Trench in the Pacific, the depth of the seabed dropped to 11,040 meters.

The *Challenger* expedition finally changed forever the idea that the ocean was a cold, lifeless, static abyss with a flat and featureless floor. Equally important, two more pieces of the plate tectonics jigsaw puzzle were now falling into place: the *mountain ranges* and the *trenches* located on the floor of the deep ocean. At the time of the *Challenger* expedition, these were interesting findings about an undersea world that had been practically unknown. And at first glance, it would seem that was all they were—interesting topographic features. No one thought that they might be linked together as part of the overriding global process that shaped the surface of the planet.

The first hint that this might be the case was supplied by Frank Taylor at the turn of the century.

A ROLE FOR THE MID-OCEAN RIDGES

One of the most far-reaching developments along the road to the theory of plate tectonics was the result of the work of the American geologist Frank Taylor (1860–1939). In 1908, Taylor proposed that the great mountain chains of the world could only have been formed by enormous *lateral* pressure, rather than by any kind of "uplift." He believed that this lateral pressure had been caused by the disintegration of a single ancient supercontinent that had broken into fragments. The fragments had then moved and collided with each other, and under the great pressures involved mountain ranges had been thrown up. Then, in a flash of inspiration that I believe should guarantee his all-time place in the history of plate tectonics, Taylor suggested that as part of this process, the Mid-Atlantic Ridge was in fact a line of rifting between the landmasses on either side of the Atlantic Ocean—that is, it was the site where the continents started to split apart.

▲

The second half of the nineteenth century was important in earth science. It ended with the work of Frank Taylor, which may be regarded as a direct precursor of the important theories of continental drift and subsequently of seafloor spreading and plate tectonics.

It's an interesting mirror on scientific research, however, that Taylor's work was not part of a linear progression toward the holy grail of plate tectonics. On the contrary, Taylor was largely ignored by his fellow scientists, and continental drift, the first of the great planet-forming theories to emerge in the twentieth century, was to be associated almost exclusively with the German geologist Alfred Wegener. In one sense, it could be said that Taylor was his own worst enemy because of the mechanism he invoked to explain the large-scale breakup of his continental mass. His proposal was that the moon was only captured by the earth during the Cretaceous period, and that as this happened, it approached closer to the earth than its present position and generated tidal currents powerful enough to drag the continents into movement. This explanation was, however, ridiculed by other scientists.

However, Taylor's ideas are of paramount interest in the historical development of plate tectonics because, for the first time, the Mid-Atlantic Ridge was identified as a zone of rifting between Africa and America; the place where the continents started to split apart. In this way, the *site* where the movement of blocks of lithosphere started, if not the *mechanism* behind it, had been pinpointed.

Taylor was a visionary who produced another piece of the plate tectonics jigsaw, but his work was almost forgotten because it was overtaken by two important ideas hovering in the wings of the new century. These were the recently proposed theory of isostasy and the giant concept of continental drift. Both ideas were to dominate the way the next generation of geologists viewed the earth.

5
ON THE ROAD TO EXPLAINING VOLCANOES, EARTHQUAKES, AND TSUNAMIS

The period from the middle of the nineteenth century, when geology first began to take the concept of "continental drift" seriously, to the middle of the twentieth century, when plate tectonics was about to flower, was one of great importance for earth science. But this was not just because of the development of great theories. Many branches of geology made considerable advances during this period. We have already explored mythical and religious explanations for volcanoes, earthquakes, and tsunamis. Now, in the light of the "new geology," we have to ask if any developments took place in the science underpinning volcanoes, earthquakes, and tsunamis in this period spanning the middle of the nineteenth century to the middle of the twentieth century—a period which marks a transition in humanity's attempts to explain the makeup of the world.

VOLCANOES

A landmark event in the study of volcanoes occurred with the eruption of Mount Pelée in 1902 (see Panel 7), when the phenomenon of *nuée ardente,* later known as pyroclastic flows, was observed. These high-temperature, dense clouds consisting of gases and small particles of lava are among the most devastating volcanic hazards.

PANEL 7
MOUNT PELÉE*
(MARTINIQUE, CARIBBEAN SEA, 1902)

"Nueé ardente, the death cloud"

The 1902 eruption of Mount Pelée—a convergent plate boundary, subduction zone stratovolcano—was of the Pelean type: A dome prevented the escape of magma, causing a pressure buildup, and large quantities of volcanic ash, dust, gas, and lava erupted and swept down the mountain as a hot avalanche, lubricated by gases— termed a "pyroclastic cloud" or nuée ardente ("glowing cloud").

Death toll—approximately 29,000.

Principal cause of death—thermal shock from pyroclastic clouds.

The Antilles Island Arc lies on the boundary of the Atlantic and Caribbean tectonic plates. It is a zone of subduction, and the silica-poor, water-rich lava produced there leads to violent volcanic eruptions.

The elegant colonial town of St. Pierre, in the prosperous French Overseas Department of Martinique in the Caribbean, lay directly under the volcano of Mount Pelée (Bald Mountain), which rises to a height of 1,400 meters. Mount Pelée is a cone-shaped stratovolcano made up of layers of volcanic ash and hardened lava, with two craters. It had been dormant since at least 1851, but when it erupted in 1902 it became the single most catastrophic volcanic disaster of the twentieth century, responsible for the deaths of 29,000 people.

Night had fallen, and the inhabitants of St. Pierre were sleeping. A watchkeeper on one of the ships in the harbor would have seen wonders as the gigantic pyrotechnic display unfurled in the night sky. Fire and smoke and clouds of ash were tossed around and the darkness was torn apart as fire danced above the mountain, and the air crackled with incandescent streaks like the trails of shooting stars. But that was merely a prelude, and just before 8 a.m., the watchkeeper would have seen the climax of the show as the mountain finally tore apart in a massive double explosion.

A black cloud was released, made up of superheated gas and incandescent volcanic ash and rock. The searing temperature within the pyroclastic cloud reached over 1,000°C. Then it split into two. One cloud shot up into the air to hover over the mountain in a great mushroom-shaped mass. But its twin was the "death cloud"—carrier of destruction as it crashed horizontally through a V-shaped notch in the side of the volcano and flowed downhill at speeds reaching 1,600 kilometers per hour.

The pyroclastic cloud, or nueé ardente, headed straight at St. Pierre. It kept close to the ground as it roared toward the town, and when the cloud hit, less than a minute after it was generated, the effect was catastrophic. The force of the cloud was enough to drag a three-ton statue off its mount and toss it aside. But it was the heat that had the most immediate impact. A terrible heat that left many fires behind as it advanced. The watchkeeper would have seen the glowing red cloud coming. He would have heard a wind blowing and seen people desperately seeking shelter where none existed. He might have thought himself safe in the waters of the harbor. But then he would have seen the pyroclastic cloud enter the sea and continue to flow above the surface of the water as a wave of fire that hit the ships. More than twenty ships burst into flame and were destroyed, and as the cloud approached his ship the watchkeeper would have seen the surface of the water torn into fiery whirlpools. And that would have been the last thing he saw. The last image impacted on his mind would have been the fiery maelstrom of the harbor, with the town burning under the mountain.

Within minutes of the eruption, St. Pierre was turned into a place of the dead, where most people were killed by inhaling the scorching fumes and burning ash. Few of the 20,000 inhabitants survived; estimates vary from one to no more than three or four survivors.

*The classification of the volcanoes, earthquakes, and tsunamis in the panels is based on post–plate tectonic thinking, and the terms used are described in the text.

The study of volcanoes changed after the eruption of Mount Pelée, when the science of volcanology entered a phase of gathering data from specific volcanoes. In this respect, Frank Perret (1867–1943) was one of the great pioneers of volcanology. After experience with the Volcanological Laboratory on Mount Vesuvius in Italy, he was involved in setting up an observatory on Hawaii in 1912. In one amazing incident, in an eruption sequence on Mount Pelée that started in 1929, Perret built a small wooden hut from which he could observe pyroclastic flows at close hand—almost too close, as it turned out, when the hut was caught in the direct path of a flow.

Volcanoes are one the most visible of all geological phenomena, and they are responsible for some of the most important features on the surface of the planet, including volcanic island arcs, huge lava plateaus, and, of course, mountains. But these are not ordinary mountains formed by folding or crumpling of the crust; they are built up from the volcano's own products.

Volcanic events have occurred throughout geological time and on the basis of size have varied from relatively small cone eruptions to what are now called *supervolcanoes*. At present, somewhere around 550 active volcanoes are found on the planet, and the United States Geological Survey (USGS) has estimated that on any one day about ten volcanoes are in the eruptive phase. These eruptions vary considerably in their degree of "explosiveness," and as a general rule smaller volcanoes are more common than larger ones.

Although they often form mountains, in the simplest terms a volcano is essentially a weak spot, or rupture, in the earth's crust through which magma, molten rock from the mantle, erupts. This magma contains gases and volatile constituents. Below the crust these gases remain trapped, but as the molten rock rises toward the surface they are released, often with explosive force. As a result, by the time it has erupted, the magma has lost much of its gaseous components and the material emitted at the surface is termed *lava*. The eruptions can take place on dry land (subaerial), underwater (submarine), or under ice (subglacial). Volcanic landscapes can be dominated by *scoria*, or cinder, fragments, which are irregular-shaped pieces of frothy lava; if the material contains gas bubble cavities, it is termed *pumice*.

VOLCANIC ROCKS

Igneous rocks can be graded from acidic to basic on the amount of silica (SiO_2) they contain:

- Acidic igneous rocks contain a high silica content (more than 63 percent). Plutonic varieties include *granite*, and volcanic varieties include *rhyolite*.

- Intermediate igneous rocks contain between 52 percent and 63 percent silica. Plutonic varieties include *diorite*, and volcanic varieties include *andesite*.

- Basic igneous rocks have a relatively low silica content (45 percent to 52 percent). Plutonic varieties include *gabbro*, and volcanic varieties include *basalt*.

- Ultrabasic igneous rocks have less than 45 percent silica. Plutonic varieties include *peridotite*, and volcanic varieties include *komatiite*.

Silica is an extremely important constituent of volcanic rocks, especially with respect to the way in which magma flows. The viscosity of magma and its ability to resist flow is partially a function of factors such as internal friction and temperature. Silicon-oxygen bonds in the form of tetrahedral begin to link together (polymerize) into long chains while the magma is still semimolten, which increases the resistance to flow. The more silica-rich a lava, such as rhyolite, the lower its temperature and the greater its viscosity— with the result that it congeals relatively quickly. In contrast, a more basic (silica-poor) lava, such as basalt, has a higher temperature and a lower viscosity, so it can flow over longer distances and is able to cover the area surrounding the volcanic vent in sheets of lava—for example, as flood basalts. An important consequence of this is that the more viscous lavas are more difficult to force through a vent, so pressure builds up that can result in explosive eruptions. The rule of thumb is, *the more silica, the stiffer the lava; the less silica, the more easily the lava flows.*

The temperature and composition of the parent magma, and the amount and kind of dissolved gases that were originally present, play an important role in setting the type of eruption that occurs. Some volcanoes erupt violently (explosive), whereas others are considerably less violent (effusive). The type of volcanic eruption depends to a large extent on the chemical composition and the viscosity of the magma—that is, on whether it is "runny" or "sticky." In general, the more runny a magma, the more easily gases can escape. In a sticky magma, the gases remain trapped, leading to a buildup in pressure and an increased risk of explosion. Runny magma therefore escapes a volcano as lava flows, whereas sticky magma is blasted into the air as pyroclastics, clouds of gas, and incandescent ash.

THE SHAPE AND MORPHOLOGY OF VOLCANOES

The different volcanic morphologies include shield volcanoes, cinder cone volcanoes, and composite cone volcanoes (or stratovolcanoes).

▸ *Shield volcanoes* have a broad, domed structure and are formed usually from basaltic lava flows.

▸ *Cone volcanoes* have one of the landscapes most associated with volcanic activity in the public perception, especially the *cinder cone* that is formed when there is a sufficient supply of pyroclastic material (i.e., lava fragments ejected during volcanic activity) to build up a cone structure.

▸ *Composite cones,* or *stratovolcanoes,* are almost symmetric, extremely violent volcanoes built up of lava flows that are emitted from a central vent and have layers of pyroclastic deposits between the flows. They can erupt over hundreds of thousands of years and provide some of the world's most spectacular mountains. They are the typical volcanoes of the Pacific Ring of Fire (Chapter 13) and include Mount Fuji (Japan), Mount St. Helens (United States), and Mount Cotopaxi (Ecuador).

Cone and composite cone volcanoes usually have a crater, a structure formed when the volcanic vent is widened by explosions and slumping. When a crater is large enough, it is termed a *caldera*—some of which can be several tens of miles across. During periods of volcanic quiescence, the craters or calderas can fill with water to form volcanic lakes. Some explosive calderas can give rise to violent eruptions.

VOLCANIC ERUPTIONS

The eruptive style of volcanoes has always caught the public imagination. In his seminal 1944 book *Principles of Physical Geology*, Arthur Holmes identified a number of principal phases in the eruptive styles of volcanic activity (see Figure 5-1). Some of these eruptive types can occur at different times in the same volcano as it evolves.

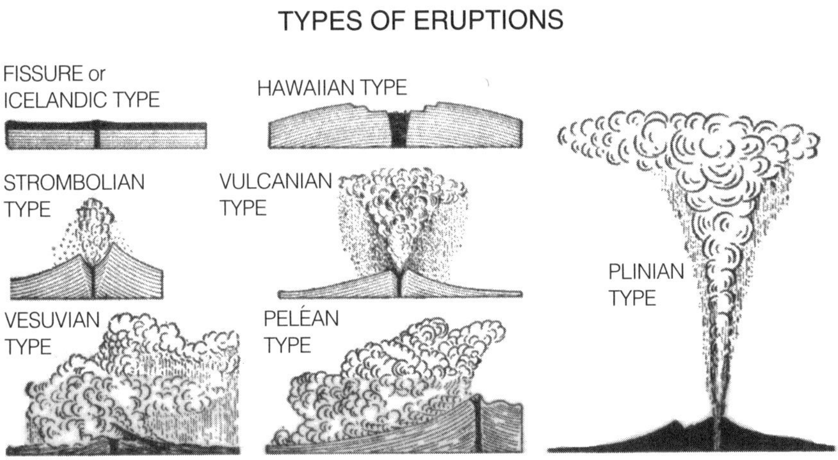

TYPES OF ERUPTIONS

FISSURE or ICELANDIC TYPE

HAWAIIAN TYPE

STROMBOLIAN TYPE

VULCANIAN TYPE

PLINIAN TYPE

VESUVIAN TYPE

PELÉAN TYPE

Figure 5-1. A diagrammatic illustration of the chief types of volcanic eruptions (from *Principles of Physical Geology*, by Arthur Holmes). Credit: Alistair Duff.

▸ *Fissure Type.* These eruptions are characteristic of the formation of basaltic flood basalts, which can spread lava over many miles surrounding the vent.

▸ *Hawaiian Type.* This eruptive style is characterized by generally quiet eruptions of mobile lava. Lava lakes can be formed, from which fountains and jets may be spewed.

▸ *Strombolian Type.* In this type of eruption, less mobile lava is exposed in a crater and trapped gases escape, leading to moderate explosions and the expulsion of ejecta. Luminous clouds may form in periods of intense activity.

▸ *Vulcanian Type.* Here, the lava is more viscous and forms a crust quickly. Gases are trapped below the lava cover and give rise to violent explosive activity. The volcanic clouds that result take on a "cauliflower" shape as they expand.

▸ *Vesuvian Type.* These are extremely violent eruptions of magma highly charged with gases built up over long periods of quiescence. Large, luminous, cauliflower-shaped clouds are formed that rise to great heights and shower down ashes.

▸ *Plinian Type.* An extremely violent form of Vesuvian eruption often culminates in a stupendous blast, throwing vapor and gases up to several miles in the air and producing a cloud. As that cloud collapses, it yields a raging pyroclastic surge and equally lethal mudflows when ash mixes with stream waters. All this makes a plinian eruption one of the great natural disasters of the world.

▸ *Pelean Type.* Sometimes highly viscous magma is prevented from escaping because a dome has formed above the escape path. Then, some magma is released through cracks, resulting in downslope avalanches (*nuées ardentes*) of gas-charged fragmental lava, which can have catastrophic effects on the population.

To measure the intensity of volcanic eruptions, the Volcanic Explosivity Index (VEI) was proposed by Chris Newell and Steve Self in 1962. Account is taken of the height to which a volcanic cloud is erupted, the volume of products erupted, and a variety of qualitative parameters, to assess the VEI of a specific volcano. Like the Richter scale for the measurement of earthquake magnitude, the VEI is a logarithmic index. A VEI of 0 is assigned to nonexplosive eruptions, and the most massive volcanoes in history are given a VEI of 8; these include plinian eruptions, such as Tambora, and some explosive caldera events. One of these was the Yellowstone Caldera VEI 8 which released 1000 cubic kilometers of material when it erupted around 650,000 years ago and covered the whole of the United States with a one-meter-thick blanket of ash. To put this in context, Mt. St. Helens ejected around one cubic kilometer in the 1980 eruption and had a VEI of 5.

THE DISTRIBUTION OF VOLCANOES

Well before the advent of plate tectonics, it was known that both active and extinct volcanoes are not distributed about the surface of the planet in a random way, but tend to lie in specific belts (see Figure 5-2), such as the Alpine-Himalayan and the Circum-Pacific belts. Up to this point, no satisfactory single explanation had been advocated to explain either the generation or the distribution of volcanoes, but it turned out that one of the most important clues lay in the way they are grouped into these activity belts.

EARTHQUAKES

Around the middle of the nineteenth century, the Dublin-born civil engineer Robert Mallet (1810–1881) carried out experiments to determine the velocity of seismic waves using gunpowder explosions. Scientists later used findings from his experiments to develop improved methods of measuring earthquakes. In terms of changing currently accepted wisdom, however, the San Francisco earthquake of 1906 was as much a watershed as the Lisbon event of 1755 had been.

Up to the end of the nineteenth century, a considerable amount of seismic data had been gathered on earthquakes and an important connection had been made between earthquakes and faults. But the nature of the connection was turned on its head following

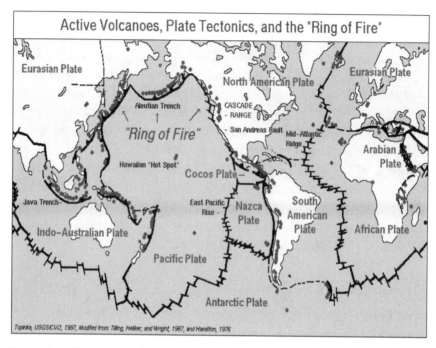

Figure 5-2. The global distribution of volcanoes. Each dot represents the site of a recently active (on a geological timescale) volcano. Three features in the distributions of volcanoes are particularly significant. 1) An important band of volcanic activity is found below sea level at the mid-ocean ridge spreading centers. 2) There is a heavy concentration of above-sea-level volcanoes in the Circum-Pacific *Ring of Fire*. 3) Some volcanic activity appears to be isolated—such as in the Hawaiian Islands that lie over an intraplate mantle "hot spot." Credit: United States Geological Survey.

the San Francisco earthquake (see Panel 8), an event that was associated with the San Andreas Fault. It was known that earthquakes generally occur around a fault line, and originally many geoscientists had thought that earthquakes generated faults. That was until Fielding Reid, who was professor of geology at Johns Hopkins University, became head of a commission to study the San Francisco earthquake and came up with his brainchild, the *elastic rebound theory*, which changed the way we looked at earthquakes. In this revolutionary theory, Reid proposed that, in fact, earthquakes were initiated by faults and not the other way around.

PANEL 8
THE SAN FRANCISCO EARTHQUAKE*
(CALIFORNIA, 1906)

"One of the costliest of all earthquakes"

Earthquake classification—"Ring of Fire," transform fault plate margin.

Magnitude—estimated at 7.8.

Death Toll—approximately 3,000.

Principal cause of death—earthquake damage and fire.

Property loss—estimated at $524 million.

▲

The San Andreas Fault was, and indeed still is, a "disaster waiting to happen." Situated on the American West Coast, it is a transform fault that lies at the boundary between two tectonic plates, the Pacific Plate on the west and the North American Plate on the east. At the fault, the Pacific Plate moves to the northwest relative to the North American Plate, causing a zone of faulting and seismic activity that's 1,300 kilometers long, up to 1.5 kilometers wide, and extends 15 kilometers into the earth. Blocks on either side of the San Andreas Fault slide past each other horizontally in segments up to 100–200 kilometers long, so it is a transform fault. The fault was born 15 million to 20 million years ago, and over that time an accumulated displacement of some 550 kilometers has taken place. The two plate boundaries involved in setting up a transform fault cannot simply slide past each other without any reaction. The friction involved creates strain, and it is the release of this strain that gives rise to earthquakes. Two kinds of movement take place across the San Andreas Fault: 1) a slow, almost constant, creep motion leading to a gradual displacement, and 2) a violent release that follows the locking up of energy for long periods until ruptures cause earthquakes (i.e., alternating periods of quiescence and violence).

At 5:12 a.m. on April 18, 1906, the San Francisco Bay area felt a foreshock. It was followed twenty to thirty seconds later by the earthquake, which had an epicenter near to San Francisco itself. Although there were many aftershocks, some quite severe, the actual earthquake itself (the shaking duration) lasted for about a minute.

It was a minute of total destruction. The earthquake was exceptional, not for the loss of life it was responsible for, which was tragic enough in itself, but for the catastrophic damage it caused in San Francisco. It was perhaps the greatest earthquake ever to rend destruction on a modern city.

The fault ruptured along a total length of some 430 kilometers, the most lengthy ever seen on the U.S. mainland, and during the earthquake roads and fences that crossed the fault were offset with a horizontal displacement of up to 10.5 kilometers. Within minutes of the main earthquake, the city was a scene of complete devastation. Road surfaces were distorted into waves and ridges several feet high, or into deep troughs, and cracks opened up. The ravaged streets were piled high with rubble from collapsed buildings and wet with water from ruptured mains. And dead people and animals were lying all around the devastated city. In what must have seemed just a moment, a prosperous, thriving city had been reduced to chaos. The problems that faced the authorities were enormous: fire, looting, people trapped in damaged buildings, hospitals destroyed. The residents of the city were fortunate, if that's the right word under the circumstances, that the U.S. military had a strong presence in the area. The military played a large role in bringing in essentials, such as food, drinking water, blankets, and tents, to the stricken city over the days following the earthquake.

One of the most terrifying aspects of the 1906 San Francisco earthquake was, in addition to the shaking damage, the extensive fires that followed the initial shock. Despite the best efforts of the firefighters, who were hampered by disruption to the city's water system, the fires spread quickly. With little alternative, the emergency services had to resort to dynamiting buildings to stop the fires from spreading. But the flames engulfed large areas within the city and survivors gathered where they could, trying to live through the long inferno of the night.

The San Francisco earthquake of 1906 was probably the most photographed in history, and several picture archives exist. The photographs in them, together with eyewitness accounts, graphically capture the horror that hit the city. Photographs of the time show raging fires, dead horses lying by their carts, twisted and split roads, firefighters struggling against overwhelming odds, and tents in makeshift camps on the nearby beaches and in the Golden Gate Park. Eyewitnesses spoke of "the swaying of the ground, of buildings crumbling, of the air being full of falling stones, of blazing fires spreading everywhere, of people trapped in the wreckage of the city, and of the heroic efforts of the police and firefighters to cope with the situation." One account told of the scenes of pandemonium at the ferry, where people were attempting to flee the city and men, women, and children fought to get aboard the vessels.

The San Francisco earthquake was so devastating because the epicenter was close to an area of high population. In all, approximately 250,000 people were left homeless out of a population of 410,000.

To leave the last word with the author Jack London, *"From every side came the roaring of flames, the crashing of walls, and the detonations of dynamite."*

*The classification of the volcanoes, earthquakes, and tsunamis in the panels is based on post–plate tectonic thinking, and the terms used are described in the text.

By studying ground displacement induced along the San Andreas Fault by the 1906 San Francisco earthquake, Reid concluded that the earthquake must have involved an "elastic rebound" of previously stored elastic strain. This elastic rebound can be explained in terms of the buildup in a fault. The rocks on either side of a fault gradually accumulate and store elastic strain energy as the ground is stretched. When the shearing stress induced in the rock exceeds the shear strength of the rock itself, the frictional resistance holding the rocks together is overcome and slippage occurs at the weakest point, until the strain is released and rupture occurs. At this stage there is an elastic rebound of the energy released as the rock attempts to spring back to its original shape, giving rise to an earthquake.

Another event at around this time that changed attitudes toward earthquakes was the Messina Strait earthquake of 1908 (see Panel 9), which brought home a severe lesson in building practice. In the earthquake, the city of Messina in southern Italy had many of its buildings destroyed, and in the subsequent rebuilding precautions were enacted to mitigate against future earthquakes.

The study of earthquakes continued into the twentieth century, and again a number of advances were generated from studies of specific earthquakes and, in particular, the seismological work of Japanese scientists. Japan and its surrounding seas are home to around 10 percent of all the world's large earthquakes (i.e., those of magnitudes greater than 6 or 7), and Japan has been called the "land of earthquakes."

PANEL 9
THE MESSINA STRAIT EARTHQUAKE*
(SICILY, 1908)

"The most powerful earthquake to hit Europe"

Earthquake classification—convergent continental /continental plate margin.

Magnitude—estimated to be 7.5.

Death toll—100,000 to 200,000.

Principal cause of death—direct earthquake damage from collapsed buildings, fire, and subsequent tsunami.

Property loss—many buildings destroyed.

▲

Sicily is in the Mediterranean earthquake band. It lies at the foot of Italy, where the Eurasian and African tectonic plates collide, and has a history of earthquake activity; for example, 30,000 people died there in the 1783 earthquake.

At the time of the 1908 event, the population of the island was almost 4 million. The earthquake struck at around 5:20 in the morning on December 28, when one, and possibly two, earthquake shocks, lasting for more than thirty seconds, were felt. The epicenter was shallow, approximately 8 kilometers below the surface in the Messina Strait that separates Sicily from the Italian mainland. The main shocks were followed by a series of aftershocks, and the earthquake struck all over southern Italy—with devastating effects in Sicily. Estimates of the dead in the port city of Messina alone ranged from 50,000 to 85,000—between one-third and one-half of the entire population of Italy's eighth largest port. Many of the buildings, which were three to five stories high, were brought down in the initial shocks, including eighty-seven of the ninety-one churches. In the aftermath, with most of the infrastructure destroyed, the survivors were left without power and water in a city reduced to rubble. To add to the chaos, water flooded into the town from broken dams.

The Italian government took time to respond to the crisis, and it was only on January 11, 1909, that martial law was enforced in the city. One macabre side of the earthquake was that many people were trapped in the rubble for days, and stories of cannibalism emerged. In other incidents, convicts freed as the jail collapsed were reported to have robbed the corpses. As time went on, Messina became known as the "city of the dead."

In the surrounding areas, cities such as Reggio Calabria suffered extensive earthquake damage, leaving large numbers of the population homeless. Many governments gave aid to the survivors, and the Red Cross alone provided $1 million in aid.

Later, it turned out that many of the buildings in the towns and cities affected had been built without any thought to earthquake damage, despite the seismic history of the region. After 1908, the situation was different and precautions against future earthquakes were taken. But for many years, thousands of people were living in temporary accommodations.

*The classification of the volcanoes, earthquakes, and tsunamis in the panels is based on post–plate tectonic thinking, and the terms used are described in the text.

The first accurate seismograph was probably developed in 1880 by the English geologist John Milne, who worked with Sir James Alfred Ewing and Thomas Grey on earthquakes in Japan. Milne is another figure who has been credited as the "father of seismology." The three British scientists were able to relate the wave patterns to the nature of an earthquake—a breakthrough in seismology.

After the Meiji Restoration in 1868, Japanese scientists put a considerable research effort into the study of earthquakes and were responsible for a number of other important advances in seismology. In one, the connection between earthquakes and faults was strengthened by the suggestion that the 1883 Koto earthquake was caused by faulting. In another, it was proposed that the 1896 Sanriku tsunami was not a result of the oscillation of bay water, but of vertical movements of the crust on the seabed.

It was perhaps the Great Kanto earthquake of 1923 (Panel 10) that had the most profound influence on seismic studies in Japan. In the wake of the most disastrous earthquake in Japanese history, in which more than 100,000 people died, a number of initiatives were implemented and earthquake institutes were set up. As the science progressed, data gathered from seismic networks were used to devise seismic wave theories, and various other mathematical studies were carried out on earthquake occurrences and seismic wave travel times.

PANEL 10
THE GREAT KANTO EARTHQUAKE*
(JAPAN, 1923)

"The most devastating earthquake to hit Japan"

Earthquake classification—convergent oceanic/continental plate margin.

Magnitude—estimated to be 7.5 to 8.5.

Death toll—100,000 to 150,000.

Principal cause of death—direct earthquake damage from collapsed buildings, fire, and subsequent tsunami.

Property loss—buildings destroyed; more than 3 million people lost homes.

▲

Japan, which lies on the Pacific Ring of Fire, is one of the most earthquake-susceptible countries on earth. It lies on a subduction zone in an area where four major tectonic plates collide: the North American, the Pacific, the Eurasian, and the Philippine plates. In this complex geological mix, the Philippine Sea and the Pacific Plate are being subducted under the Eurasian and North American plates, which are themselves colliding. All this leads to an extremely unstable tectonic environment rent by volcanic and earthquake activity. In fact, the Japanese islands have more earthquakes (several thousand a year, if very small tremors are included) than any other country—with destructive earthquakes happening about once every two years.

The Great Kanto earthquake hit the Kanto plain, an area with an estimated population of around 2 million, on the island of Honshu, two minutes before noon on September 1, 1923, with an epicenter lying beneath Sagami Bay. The damage was widespread, especially to the cities of Tokyo and Yokohama, which were both left in ruins. The death toll was immense and was put as high as 150,000 people, according to some estimates.

The damage took many forms. Some damage occurred directly through shock waves causing movement of the ground, and in one reported incident the statue of the Great Buddha at Kamakura was rocked on its foundations. Some effects were magnified because of local conditions; for example, Japanese homes are generally lightweight structures of wood and paper and would not have suffered much damage as they collapsed. The houses were, however, extremely susceptible to fire—a danger that was made considerably worse by the fact that the earthquake struck at noon when the midday meal was being prepared on charcoal stoves. The fire hazard was also increased by the presence of inflammable industrial materials and munitions in the area.

The first shocks from the earthquake are reported to have lasted from four to ten minutes, and fires from the stoves broke out immediately, spreading rapidly as a high wind drove the flames through the close-packed dwellings and created vast firestorms. One fire killed as many as 30,000 people seeking shelter in an open space. Matters became more desperate as the shock waves fractured water mains, making the job of firefighters almost impossible. In total, around half a million homes were destroyed.

Most victims of the earthquake died either from being burned in the fires or from suffocation, as the raging flames destroyed around 80 percent of the cities of Tokyo and Yoko-hama. In addition, deaths occurred from the subsequent tsu-namis.

Another consequence of the earthquake was that it set off large landslides in the hilly areas around the cities. These landslides changed the topography of the countryside and hurled trains, and sometimes entire villages, into the sea.

The Great Kanto disaster suffered the deadly three-prong attack from earthquake, fire, and tsunami, and it was one of the greatest natural disasters of all time. It was not until the early 1930s that Tokyo and Yokohama were totally rebuilt.

*The classification of the volcanoes, earthquakes, and tsunamis in the panels is based on post–plate tectonic thinking, and the terms used are described in the text.

Studies such as those carried out in the United States and Japan led to a greater understanding of earthquakes, and a number of important seismic parameters could now be identified.

Forces that act on rocks can produce various stresses that lead to the rocks being put under strain. If the rocks are strained to breaking point, the strain energy can be released in a sudden rupture that generates shock vibrations in the form of seismic waves. In this context, an earthquake can be defined as *the passage of vibrations through rocks, caused by the rapid release of accumulated strain energy.* This energy can, if the shock is great enough, generate waves in a series of terrifying ground undulations.

Earthquakes can therefore be thought of as providing a natural mechanism for the planet to relieve the stress building up in it. One particularly dangerous characteristic of earthquakes, however, is that all the potential strain is not necessarily released during the main shock, and this may result in aftershocks occurring.

At this stage, it is necessary to define a few of the terms used by geophysicists when they describe earthquakes. The vibrations resulting from the release of strain by an earthquake travel through rocks in all directions from a *focal point* (or focus) where the energy is released; this is the point on a rupturing fault where the first movement begins, and it is located at a depth termed the *focal depth*. Another name for the focal point is the hypocenter (meaning "below the center"). The *epicenter* of an earthquake is the point on the ground directly above the focal point. The vibrations generated by an earthquake are in the form of seismic waves. Theoretically, waves from an earthquake can be detected throughout the planet, although in practice the energy falls off rapidly and the waves grow weaker as the distance from the focus increases.

EARTHQUAKE DETECTION AND MEASUREMENT

A number of different kinds of seismic waves are generated during earthquake activity and they are recorded as lines on a seismograph. One set of waves travels through the *surface* of the earth, and one travels around the *body* of the earth.

L-waves, or Love waves (named after the British physicist A.E.H. Love), and Rayleigh waves (named after the British physicist Lord Rayleigh) are surface waves that are confined to the crust. Surface waves are long slow waves that produce a rocking sensation: Rayleigh waves make the ground roll, whereas Love waves travel horizontally and make the ground move from side to side.

Body waves are transmitted through the interior of the earth and are divided into P-waves and S-waves.

P-waves (primary waves) are "push and pull," or compressional waves in which particles vibrate in the direction in which the wave propagates. They are the fastest waves, with speeds of about 6 kilometers per second in granite, and they are the first waves detected by monitoring instruments. Importantly, P-waves can be transmitted through both fluids and solid rock. They cause an abrupt shock and make structures contract and expand.

S-waves (secondary waves) are waves in which particles vibrate at right angles to the wave propagation, making rocks move from side to side. S-waves exhibit a periodic motion that causes structures to shake and is responsible for extensive earthquake damage. As S-waves are slower (4 kilometers per second in granite), they follow after P-waves, but they can only travel through solid rock and not through a liquid medium. In practice, the actual speeds of S-waves and P-waves depend on the elasticity and density of the material they are transmitted through.

Earthquake waves are detected and recorded using seismographs—instruments designed to measure the motion of the ground during, and after, an earthquake. Since they were first invented, seismographs have undergone considerable modification and have evolved into the present-day electronic versions, in which electromagnetic recorders are usually connected to computers with analog-to-digital converters. In this kind of instrument, a voltage is generated as the ground shakes and is then amplified and recorded. When they are positioned in suitable arrays, seismographs can precisely locate the source of an earthquake. For example, if the arrival times of P-waves and S-waves are known at three stations, the location of the epicenter of the earthquake, and the time at which the event occurred, can be determined by triangulation.

Magnitude, intensity, and depth of focus data are three different ways of measuring earthquakes.

Magnitude. The size, or strength, of an earthquake can be assessed in terms of its magnitude as measured on a seismograph. The magnitude of a specific earthquake depends, among other factors, on the size of the fault involved in the generation of the earthquake, the nature of the rocks involved, and the amount of stress that has built up. One of the most widely used magnitude scales is the *Richter scale,* named after Charles F. Richter of the California Institute of Technology who introduced it in 1935. Richter recognized that when seismic waves generated by an earthquake are recorded, they offer an estimate of the magnitude of the earthquake. The Richter scale is based on the instrumental measurement of the *amplitude* of seismic waves, which is related to the amount of energy released in an earthquake. It was originally based on measurements recorded using a particular instrument—the Wood-Anderson torsion seismometer, which is usually referred to as the M_L, or local magnitude, scale. The Richter scale is a log-scale so that, for example, an earthquake with a magnitude of 4 on the scale has ten times the *intensity* of one with a magnitude of 3. This does not mean that the earthquake is ten times *stronger,* because the amount of energy released in an earthquake is increased by a factor of approximately 32 for every tenfold increase in amplitude. On the Richter scale, an earthquake with a magnitude of 2 is the smallest felt by humans.

There are a number of other ways of classifying earthquakes on the basis of their magnitude, most of which are modifications of the Richter scale. To help avoid the confusion, the USGS advises a simplified approach in which the term *magnitude* and the symbol *M* (without subscripts or superscripts) is used to describe earthquakes, with additional information provided only if requested. We will use this system in our treatment of earthquakes.

Intensity. This is a semiquantitative measure of the effect that ground shaking, caused by an earthquake, has at a specific location. One way of gauging intensity is to assess the damage the ground shaking does to both humans and the environment during an earthquake event. The Mercalli scale, devised in 1902 and later modified, is designed to do just this. The scale now has twelve divisions ranging from I ("not felt, except by a very few") to XII ("waves seen on ground, damage total"). It must be remembered, however, that the Mercalli scale relates to the intensity of an earthquake at a particular place, and as the effects will fall off with distance away from the epicenter of an earthquake, it is only a relative measure of intensity.

It must be remembered, too, that magnitude and intensity are different but equally useful ways of categorizing earthquakes. Magnitude is a measure of the size of an earthquake and doesn't change, whereas intensity is a measure of the amount of ground shaking caused by an earthquake and decreases away from the epicenter.

Focal Depth. Another parameter used in the description of earthquakes is focal depth. This is the depth below the earth's surface at which the energy of an earthquake is released. There are several schemes for classifying earthquakes on the basis of depth of focus, one of which has the following three categories: shallow focus (less than 70 kilometers), intermediate focus (70–300 kilometers), and deep focus (greater than 300 kilometers). Around 90 percent of all recorded earthquakes have focal depths less than 100 kilometers.

THE DISTRIBUTION OF EARTHQUAKES

As was the case for volcanoes, it had been known for a long time that earthquakes were not distributed randomly around the planet, but were concentrated in belts. Worldwide, there were three predominant seismic belts:

1. The Circum-Pacific "Ring of Fire" belt (see Figure 5-3) is very important in volcanic activity. In total, around 80 to 90 percent of the energy released in earthquakes occurs in this belt, which is home to some of the most destructive earthquakes in the world. New Zealand is included in this belt, and it has an interesting feature in the Alpine Fault, which runs up the spine of the North Island and is the terrestrial boundary between the Pacific and Australian plates. The boundary has ruptured several times over the last thousand years and has given rise to some spectacular scenery in this seismically active region.

2. The Mediterranean-Asiatic belt, also known as the Alpide belt, stretches from the islands of Java, Bali, and Timor through the

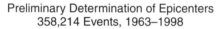

Preliminary Determination of Epicenters
358,214 Events, 1963–1998

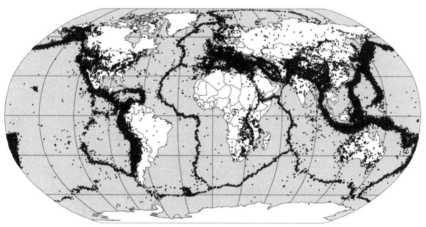

Figure 5-3. The global distribution of earthquakes. Worldwide, there are three predominant seismic belts. 1) The *Circum-Pacific "Ring of Fire" belt* is home to some of the most destructive earthquakes in the world. 2) The *Mediterranean-Asiatic,* or *Alpide, belt* stretches from the islands of Java, Bali, and Timor, through the Himalayas, the Mediterranean, and out into the Atlantic. 3) The *Mid-Ocean Ridge belt* is found at the mid-ocean ridge spreading centers, and earthquakes here tend to have a magnitude less than 5. The San Andreas Fault runs for around 1,300 kilometers inland from the northern end of the East Pacific Rise section of the Pacific Mid-Ocean Ridge system. The fault marks a transform boundary between the North American and Pacific plates.

The intraplate earthquakes are not concentrated into specific bands, but they have included some of the biggest earthquakes ever recorded, such as those occurring in eastern Asia in the interior of the Eurasian Plate (see Figure 11-1). Credit: NASA.

Himalayas, the Mediterranean, and out into the Atlantic. Around 10 to 15 percent of the world's earthquakes occur in this belt. There is some confusion over where Indonesia sits in the earthquake classification scheme, but it seems to lie between the Alpide belt and the Pacific Ring of Fire. The distribution of these belts was to prove a crucial tool in understanding plate tectonics.

3. The Mid-Ocean Ridge belt is found at the mid-ocean ridge spreading centers, and earthquakes here tend to have a magnitude of less than 5.

TSUNAMIS

The name *tsunami* comes from two Japanese words, *tsu* and *nami*, and means "harbor wave." That name may have arisen because tsunamis have almost no effect in the open ocean, but they become killers once they reach the shore. In the past, tsunamis have sometimes been referred to as tidal waves. But this is far from the truth since tidal forces are not involved in any way in the formation of tsunamis. Instead, a tsunami is a very special kind of wave that arises under a very particular set of conditions (see Figure 5-4).

Oceanographers describe waves in terms of a number of characteristics. The wave *crest* is the highest point on a wave. The wave *trough* is the lowest point on a wave. The wave *height* is the vertical distance from the top of the crest to the bottom of the trough. The *wavelength* is the distance between two successive crests or troughs. The wave *amplitude* is one half of the wave height. The wave *period* is the time taken for two successive waves, or troughs, to pass a point in space (i.e., a measure of the size of a wave in relation to time).

In general, there are three main types of waves in the oceans: sea waves (surface, mainly wind-driven waves), tidal waves (mainly gravity-driven), and tsunamis (sudden event–driven).

As the wind passes over the sea surface, it creates friction that donates energy to the water and forces the sea surface to pile up water into ripples and ridges—the surface waves. The thin upper skin of the ocean is continually in motion, and what we notice when we look out to sea are these surface waves that extend down to a relatively shallow depth of a few meters. The size and shape of these surface waves are influenced by wind speed, wind fetch

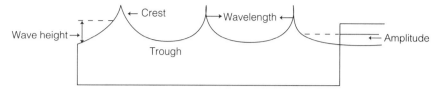

Figure 5-4. Sea wave characteristics. A number of special characteristics define a tsunami relative to wind-driven ocean waves. 1) Tsunamis have a relatively small wave height, but a very large wavelength, which may be up to hundreds of kilometers, compared to a maximum of a few hundred meters for wind-generated surface waves. 2) Tsunamis move at relatively very fast speeds—up to several hundred kilometers per hour, compared to 15 to 30 kilometers per hour for most wind-generated waves. 3) Tsunamis extend all the way down the water column, whereas wind-generated waves extend down to a maximum of a few meters.

(i.e., the distance the wind has traveled over open water—which can be large in the open ocean), and the duration of the wind gust.

When water in a wave is set in motion, all is not what it at first seems. When a wave moves below the surface, the water particles are drawn into a nearly circular orbit. But a wave is the forward motion of *energy*, not water. This means that while it is true that a wave moves with the wind, the individual water particles in their circular orbit do not. The analogy often given is that a cork in a wave would simply go round and round the same circular orbit in the same place.

What happens to surface waves during their lifetime? Remember, it is the form of the wave itself that moves, not the water. Generally, surface waves are short waves having typical wave periods of about 15 seconds and relatively slow speeds, usually in the range of less than 10 centimeters per second to greater than 25 centimeters per second. The waves can travel for thousands of miles across the ocean surface before they encounter a coastline. Then, as they move into shallow water, things begin to happen. As the wave encroaches on the coastline, there is a critical point when the water depth is one half the wavelength of a wave, and that's when changes start to occur. The wave alters shape, increases in height, and begins to slow down. The particles of water in the circular motion orbits begin to meet the seabed and the wave breaks and finally disintegrates to end its existence as a breaker.

The characteristics of a "normal" surface ocean wave can be

summarized in this way: It is generated by the wind, it is relatively short, and it moves at a relatively slow speed.

Tides are a repeated cyclic rising and falling movement of the bodies of water on the surface of the planet in response to the gravitational effects of the sun and the moon, modified by the earth's rotation. Although the sea level changes from tidal movements are most detectable on the shoreline, they operate globally and on an ocean scale. In this way, tidal waves can be thought of as very large, long waves moving across the surface of the earth over periods of several hours.

A tsunami is a series of waves of extremely long wavelength generated when a large volume of water in an ocean is shifted suddenly in a rapid displacement. Tsunamis, however, do not behave like ordinary waves.

Most tsunamis are formed when a sudden release of energy from an undersea earthquake, or a landslide associated with it, causes the water column above to be displaced and to move vertically up and down. But there are other ways in which tsunamis can be generated. One of these is by a volcanic eruption. For example, a tsunami may follow a *phreatomagmatic* explosion caused by the formation of high-pressure steam after magma mixes with water. The tsunami following the eruption and collapse of Krakatau in 1883 (see Chapter 1, Panel 3) was one of the most destructive ever recorded. One of the most spectacular tsunamis in history occurred in Lituya Bay, Alaska, in 1958, when a massive rockslide into the sea generated a great wave. Meteorite impacts on the earth's surface can also cause tsunamis. Nonetheless, most tsunamis, over 90 percent, arise because of *vertical* movement caused when an earthquake event deforms the crust and displaces the water column. Usually an earthquake with a magnitude greater than 7 on the Richter scale is required to generate a major tsunami. The water displacement happens because up-and-down movements in the crust raise or lower the water column above—as if the ocean is being "rung by a bell." The water column then attempts to regain equilibrium by producing waves.

It is easy to understand why tsunamis have in the past been confused with tidal waves, since the scale and physical characteristics of the two have a number of similarities, and tsunamis do resemble tides in some ways. However, two points must be borne in mind here. The first is that most tides are relatively gentle, and although they regularly expose and cover coastal tracks at high and low water, they usually do no harm. In contrast, tsunamis can strike the shore with devastating effect. The second point to remember is that tides can be regarded as continuous "background" events, whereas tsunamis arise because of some sudden one-off release of energy—such as that caused by earthquake activity.

Five characteristics of tsunamis are critical to understanding their behavior patterns:

▸ Tsunamis have a relatively small wave height (amplitude), usually from a few centimeters to less than a meter, but a very long wavelength that may be *hundreds of kilometers*. They also have a very long period, sometimes as much as hours, between wave crests. In contrast, wind-driven surface waves have a maximum wavelength of between 100–150 meters and a period of a few seconds.

▸ Unlike the situation for wind-driven surface waves, which are shallow and only extend down a few meters, the entire depth of the water column moves with a tsunami, regardless of how deep the water column is. As a result, tsunamis transport a tremendous amount of energy, with little loss, as they move across an ocean.

▸ The speed of both surface wind-driven waves and tsunamis is a function of wave depth, and here it is important to remember that wind-driven waves extend no more than a few meters from the ocean surface and are relatively slow. Tsunamis, by contrast, extend all the way down the water column and are extremely fast. Tsunamis travel at speeds that can reach as much as 800–950 kilometers per hour. An important consequence of these speeds is that a tsunami can cross an entire ocean in hours and can strike at places far from its source. In the 1960 Chilean tsunami, the waves reached Japan, almost 17,000 kilometers away, in under twenty-four hours (see Panel 11).

▸ Several tsunamis (a *train*) can be generated by the same event, but—and this is extremely important—the first to arrive at a coastal site may not be the most destructive.

▸ Tsunamis propagate outward from their source in *all* directions, although they may be stronger in some directions than others.

Tsunamis can therefore travel at high speed, for long periods of time over large distances, with the loss of very little energy.

PANEL 11
THE CHILEAN TSUNAMI*
(1960)

"Generated by the strongest earthquake ever recorded"

Location—South Pacific Ocean.

Tsunami classification—earthquake generated.

Earthquake magnitude—9.5.

Tectonic setting—Pacific "Ring of Fire".

Death toll—~2,200, but possibly as high as 6,000.

Principal cause of death—direct earthquake damage and tsunami.

▲

The Great Chilean Earthquake that struck on May 22, 1960, was the strongest ever recorded, with a magnitude of 9.5. It occurred along the Peru-Chile Trench as the Nazca Plate was subducted below continental South America. The earthquake triggered a tsunami that was felt throughout the Pacific.

Locally, the tsunami struck the Chile/Peru coast ten to fifteen minutes after the start of the earthquake event. Some people had already taken to boats to escape the ground shaking, only to be struck by 25-meter waves as the tsunami hit the coast and drove inland for half a kilometer. In the region closest to the epicenter, some 2,000 people were killed, and overall a total of 2 million people were left homeless by the earthquake and tsunami.

In about fifteen hours the tsunami had traveled 10,000 kilometers and reached the Hawaiian islands, where a wave 90 meters high struck the coast, killing more than fifty people. Such was the force of the waves hitting Hawaii that metal parking meters along the coast were bent over. In twenty-two hours, the tsunami had crossed the entire Pacific and hit Japan, where 3-meter waves were recorded.

The overall damage costs of the whole event was estimated at anywhere from $400 million to $800 million.

*The classification of the volcanoes, earthquakes, and tsunamis in the panels is based on post–plate tectonic thinking, and the terms used are described in the text.

A number of stages can be identified in the life of an earthquake-generated tsunami. It is born in the shaking of the seafloor around the epicenter of an earthquake. The shaking pushes the water column up and down, and the energy associated with heaving the water above its normal level is transferred to the horizontal movement of the tsunami. Tsunamis travel outward (i.e., propagate) from the point at which the trigger event occurred in all directions, much as the ripples formed on the surface of a pond develop as the water is disturbed by a stone. The further away from the trigger, the progressively weaker a tsunami becomes.

For most of its life, a tsunami is a *hidden* wave. In the open ocean, tsunamis look harmless (if they can be seen at all) and a ship could move through most of them without even being aware of it. The danger lurks because of the vast amount of water behind and below the wave front—water that stores vast amounts of energy as it moves.

As the long, low, and fast tsunami waves approach a shoreline, the wave energy begins to be compressed and a number of things happen. As a tsunami enters the relatively shallow waters of the continental shelf—usually less than 100 meters deep—both the wavelength and the speed of the wave front decrease. However, as the speed of the wave front slows down, the bulk of the tsunami behind it maintains its fast speed and, as a result, the height of the wave *increases* as the water piles up on itself. As this happens, the energy can be concentrated into a huge destructive wave, sometimes up to several hundred feet high. As a rule of thumb, the steeper the shoreline, the higher the tsunami.

A tsunami can appear in a number of ways. First, if the trough of the tsunami is the first to reach the coastline, the water can be dragged back (*drawback*), exposing parts of the shore that are usually underwater and stranding many marine creatures. Drawback can be an advance indication that a tsunami is approaching the shoreline, but the problem is that the time between the water receding and the full tsunami moving in is usually only seconds or, at best, minutes.

By far the most dramatic image of a tsunami in many people's minds is when it manifests itself as a huge wave crashing down on the beach and carrying great volumes of water inland. This is all the more terrifying when there is no warning at all and a massive wall of water suddenly appears apparently out of nowhere and thunders onto a beach. As well as this initial impact, a tsunami can drive large volumes of water onto the land (*run-up*), which can have catastrophic effects on populated areas. In reality, however, most tsunamis do not come in as massive "disaster movie" waves. Rather, they encroach on the shoreline much like very strong, very fast, very high tides would do (hence the confusion between the two types of waves) and cause damage as they rush up the beach and move inland.

Attempts have been made to produce tsunami intensity scales, similar to the Mercalli scale for earthquakes. In 1927, August Seiberg devised an intensity scale that was later modified by Nicholas Ambraseys. This 6-point scale ranged from 1, "very light," to 6, "disastrous." Then in 2001, Gerassimos Papadopoulos and Fumihiko Imamura brought out an extended twelve-point scale, with the divisions partially related to tsunami wave height. The Papadopoulos-Imamura scale ranges from 1, "not felt (by humans)," to 12, "completely devastating."

THE NEXT STAGE

This chapter opened with the question: Did any developments take place in the science underpinning volcanoes, earthquakes, and tsunamis in the period from the middle of the nineteenth century to the middle of the twentieth century? The answer, as we have seen, is yes—a great many.

To quote Arthur Holmes in *Principles of Physical Geology* (1953 edition):

> *We no longer blindly accept events as results of the unpredictable whims of mythological beings. . . . Volcanic eruptions and earthquakes no longer reflect the erratic behavior of the gods of the underworld: They arise from the action of the earth's internal heat on and through the surrounding crust. The source of the energy lies in the material of the inner earth.*

This is a very telling quote, because Holmes linked volcanoes and earthquakes together and believed that they both arose from "the action of the earth's internal heat." Although he suggested that the energy source "lies in the material of the inner earth," there was still a lack of understanding of the mechanism, or mechanisms, that actually caused volcanoes and earthquakes.

One important characteristic of both volcanoes and earthquakes is that they are not distributed at random around the surface of the globe. Rather, they are concentrated in a number of "activity belts." Jumping ahead of the game for a moment, it seems that at the time, scientists would have asked if there could be a tie-in between this "action of the earth's internal heat" and the distribution of volcanoes and earthquakes, and if the distribution patterns were the key to unlocking the secrets of these phenomena?

They were certainly important but, as it turned out, they were only part of a huge body of evidence that was to underpin the great theories that emerged in the twentieth century, especially plate tectonics. To trace the story of how these theories developed, we must return to the early years of the last century.

6
CONTINENTAL DRIFT: A
THEORY WITHOUT A CAUSE

A BIG DEBATE

The debate between those supporting the permanency of the landmasses and those advocating movement of the continents continued to rage. But there was now a considerable body of evidence that the continents had, in fact, moved apart.

- ▶ The geographic "fit" of the landmasses had been demonstrated.

- ▶ Paleographic mapping indicated the existence of past supercontinents.

- ▶ Similarities in fossil flora and fauna in continents now separated by oceans were discovered. For example, the widespread fossil remains of some dinosaurs suggested that North America, Europe, and Africa had once been connected.

- ▶ Identical belts of rocks, stratigraphic sequences, and even mountain ranges were found to exist in continents that are now separated.

- ▶ Fossil evidence indicated that the continents had had strikingly different climates in the past.

- ▶ The fact that the earliest sediments in the Atlantic are from the Jurassic age suggested that the ocean did not exist prior to this time, although much older rocks are found on the surrounding landmasses.

Although the idea was vigorously contested by those still supporting theories involving land bridges, the underlying concept was in place now—the continents had moved apart. However, a pivotal problem seized upon by opponents of moving continents was that no plausible mechanism had yet been found to drive the process. This was to remain the great stumbling block to the acceptance of the concept of the moving continents for decades to follow, and in the early part of the twentieth century the fact was that no generally acceptable explanation had been proposed to account for the movement of the landmasses.

In this climate of uncertainty, a number of important ideas on how the surface of the earth had evolved began to emerge. Prime among these were isostasy and the theory of continental drift itself.

ISOSTASY

At the beginning of the twentieth century the idea of isostasy, or gravitational equilibrium, had been around for some time under one name or another, and as far back as 1855, George Airy (1801–1892) had suggested that the earth's crust was in a state of equilibrium.

The surface of the continents is varied in the extreme and includes high mountain ranges, vast plains, and rift valleys. At the start of the twentieth century, evidence was beginning to accumulate that the floor of the oceans also had a varied topography, with mountain ranges, flat plains, and deep trenches. Although both environments had a varied surface topography, the lithosphere (the crust and upper part of the mantle) is considerably thicker under the continents than under the oceans. This is an important distinction and plays a central role in the theories that attempt to explain the nature of the forces that shape the earth's crust.

The thick layer of crust under the continents covers around 30 percent of the earth's surface and consists predominantly of what have been termed "light," less dense rocks, mainly granite, which we have already referred to as SIAL. In contrast, the thinner crust under the oceans, which comprises around 70 percent of the earth's surface, is made up chiefly of denser "heavy" rocks, mainly basalt, and is termed SIMA (see Chapter 3). In addition to differences in composition, the continental and ocean crusts have very different ages, the oceanic crust being relatively young, none of it older than around 200 million years, whereas the oldest material on the continents is more than 3.5 billion years old.

Both the continental and the oceanic crusts rest, or float, on the mantle in the same way. However, because the oceanic crust is thinner and made of more dense material, it does not stand as high and does not penetrate as deeply into the mantle as does the continental crust. The overall picture, therefore, is one of elevated continents and depressed ocean basins.

It would appear as if the differences in thickness of the major crustal blocks are balanced by how high they float on, and how deeply they penetrate into, the upper mantle. If that's true, then at the base of the crust there must be a layer that is plastic enough to be able to adjust to differences in the nature of the crustal blocks. This layer is the asthenosphere (see Chapter 2). Overall, therefore, it would seem that the lithosphere is in a state of balance, or gravitational equilibrium, with the asthenosphere. In the state of equilibrium, the heights of the continents and the ocean floors are controlled by the densities of the rocks that make them up. In 1889, the American geologist Clarence Dutton (1841–1912) explained this state of equilibrium through the formal concept of *isostasy*, in which lighter material, such as granite, rises in the mantle, whereas heavier material, such as basalt, sinks (i.e., buoyancy is effected).

The theory of isostasy, with its gravitational equilibrium between the crust and the asthenosphere and its claim of "floating" continents, was an important background against which the theory of continental drift made its appearance.

CONTINENTAL DRIFT

The ideas that began to emerge in the last few decades of the nineteenth into the early twentieth century could be said to have culminated in the theory of *continental drift*. But it must be stressed that continental drift itself wasn't strictly a new idea, but rather it was one stage on a journey that stretched back centuries. We can identify four important milestones on this journey.

The first was when the Elizabethan philosopher Francis Bacon used evidence from emerging global-scale mapping to suggest that the west coast of Africa and the east coast of South America could fit together, like pieces in a massive jigsaw puzzle. The second, around 1880, was Alexander von Humboldt's proposal that the two continents, Africa and America, had once been joined and had subsequently *drifted apart*. The third came when Antonio Snider-Pellegrini produced his famous diagram to show *how* Africa and South America may have once been joined. The fourth milestone was Frank Taylor's suggestion that the Mid-Atlantic Ridge was a *line of rifting* where Africa and America had drifted apart.

Although it is clearly unfair to others, such as Humboldt, Snider-Pellegrini, and Taylor, the one name that will forever be most closely associated with the theory of continental drift is that of Berlin-born scientist Alfred Wegener.

Wegener (1880–1930) first became interested in the evolution of the earth's crust when he read a paper on the topic of a land bridge between South America and Africa, which explained the similarity in flora and fauna in the two areas now separated by the Atlantic Ocean. But he dismissed the concept of a land bridge because, for one thing, it required the sinking of less dense granitic rock into the more dense basaltic rock of the ocean floors—which Wegener knew was clearly not feasible.

From then on Wegener was hooked, although it must be remembered that when he suggested that the continents had moved, he was not sailing into entirely uncharted waters; rather, he was drawing together many of the strands that geologists had previously identified. Nonetheless, Wegener's proposal of continental drift became one of the great turning points in earth science.

Wegener was a German meteorologist with a variety of interests. He studied natural sciences at the University of Berlin and gained a doctorate in astronomy in 1904. He then began to work for the Royal Prussian Aeronautical Observatory, and in 1906 he was a member of a Danish expedition to Greenland. On his return, he joined the teaching faculty at the University of Marburg, where he lectured on meteorology, among other topics. In 1911, his book on *The Thermodynamics of the Atmosphere* was published.

Clearly, Wegener was a multidisciplinary scientist whose background helped him to draw together the various strands that made up his theory of continental drift. Despite the massive criticism that followed the publication of the theory, Wegener was awarded a professorship at the University of Graz in Austria, and he continued to gather evidence in support of his theory until his death on an expedition to Greenland in 1930.

So what made continental drift theory such a landmark in earth science? The answer is that it was the first of the three great modern theories that attempted to explain how the surface of the earth had evolved.

There are three key elements to the theory of continental drift:

1. Around 250 million years ago, there was only one landmass on the surface of the earth—a supercontinent named Pangaea ("all lands" or "all earth") by Wegener.

2. Pangaea fragmented, forming blocks of crust that eventually became the present-day continents. In the first stage, Laurasia, the northern part of the supercontinent that became North America and Eurasia, separated from the southern part, Gondwanaland, which was to form Africa, South America, India, Australia, and Antarctica.

3. Following this initial split, the continents continued to separate and drifted to their present positions, which they reached by "ploughing through" the oceanic crust.

Wegener proposed that the continents, which are composed of lighter granitic material, moved by drifting on, or ploughing through, the denser basalt layer of the ocean floor. He also suggested that the mechanism behind the drifting was centrifugal force, created by the rotation of the earth, that drew the landmasses away from the poles (a concept called *Polflucht*) toward the equator by gravitational attraction to the equatorial bulge. Tidal forces from the sun and the moon were also thought to cause lateral movement of the continents. (The Pangaea breakup sequence is illustrated in Chapter 4, Figure 4-2.)

In 1915, Wegener published his great work, *The Origin of the Continents and Oceans*, a book that compares in importance with Darwin's *Origin of Species*. And like Darwin's seminal publication, Wegener's theory courted controversy from the moment it was first proposed in 1911. In fact, many in the scientific community rejected continental drift theory almost out of hand, and turned on Wegener himself with an outcry that sometimes verged on the vitriolic. The principal ground for rejection of the theory was that the forces Wegener had put forward to drive the movement of the continents were not strong enough to work. Wegener, however, held his nerve and continued to collect evidence in support of his theory until his death in 1930.

Why did Wegener's ideas raise such a maelstrom of dismissive criticism and an outpouring of abuse? Ridicule of the kind expressed by the president of the American Philosophical Society at the time, who described the theory as "Utter damned rot." There were equally damning comments from other scholars, including one criticism along the lines that "anyone who valued his scientific sanity would never dare support such a theory."

Fossil evidence had quite clearly shown that a number of species had spread between what were now widely separated continents. This was not in question, and in order for it to happen there must have been some form of physical contact between the landmasses. This also was not in question. The controversy raged over what *form* this contact had taken, with the two main theories being either that the continents had once been joined together as a supercontinent that had subsequently split up, or that they had been independent units connected by land bridges.

Paleographic maps, which reconstructed past conditions, suggested that the landmasses had once been joined into supercontinents. Wegener had followed up on this observation in his theory of continental drift by bringing to the forefront the notion that the continents had indeed changed their positions on the surface of the planet, after the breakup of a supercontinent.

Good science? Perhaps.

But even for the most die-hard supporters of continental drift, one massive stumbling block still remained: the mechanism behind the movement.

And it is here that I began to perhaps sense the attitude of the entrenched establishment. It wasn't simply that most geoscientists did not believe that an acceptable mechanism to move the continents had yet been identified. No, it went much deeper than that—they did not believe that continental blocks could *ever* move through the rigid basaltic crust of the oceans, whatever force was applied.

In the years following the publication of Wegener's theory, a great deal of attention was focused on identifying the mechanism that could move the continents and so allow the geophysical dynamics of continental drift to be accepted. In a way, it is deeply ironic that this search was directed toward a phenomenon that never happened. Because, as we shall see later, the continents did *not* "plough through" the rigid basalt on the floor of the oceans at all.

Nonetheless, the initial post-Wegener thinking on the mechanism for wandering continents is an important stage in the history of earth science.

IN THE SHADOW OF CONTINENTAL DRIFT

Not everyone was opposed to Wegener, and three figures in particular entered the fray on his behalf, injecting important ideas into the controversy of continental drift.

One was the English scientist Arthur Holmes (1890–1965), whose career as a geologist spanned a time when several great debates were raging in earth science. We have already seen that Holmes made a very significant contribution to the debate on the age of the earth. Later in his career, he made an equally important impact on the most controversial aspect of continental drift—namely, the mechanism that drove the movement of the landmasses around the surface of the earth.

The mechanism Holmes proposed was the existence of convection currents in the mantle that could cause large-scale movements in the crust of the earth. When Wegener proposed his theory, the earth was generally thought to be a solid spherical body with little motion occurring in the interior. But in his research into the age of the planet, Holmes had worked with radioactivity, so he knew that the radioactive elements could be a source of heat in the interior of the earth, together perhaps with heat generated by the molten core (see Chapter 2). At this stage, Holmes revealed his genius when he suggested that this internal heat would generate *convection currents* within the mantle. Although these subcrustal convection currents were thought to be slow and ponderous, they followed a pattern in which great convection cells were developed; each cell had an upwelling limb, with movement toward the base of the litho-sphere, and a down-welling limb with movement into the interior of the earth.

The cells work in this way: Hot liquid, or in this case molten rock in the interior of the earth, rises to displace cooler rock, creat-ing a current as it does so and setting up convection cells. The upper limbs of the cells start to rise and eventually touch the base of the lithosphere, where they diverge in opposite directions as the cells continue to go around. It is the effect that these subcrustal convection cells have on the surface of the earth that is important, because as the cells ascend then descend, they drag the continents into movement (see Figure 6-1). Holmes proposed that the turn-over of these convection cells could provide the mechanism behind the movement of continents.

In his famous textbook *The Principles of Physical Geology*, Holmes made what were, for the time, quite staggering suggestions when he proposed that convection currents flowing horizontally would drag the continents along with them. But since the floors of the oceans are composed of basalt, this rock must be continuously *moved out of the way* (my italics). This, Holmes concluded, is most likely to happen when two opposing cells come to the surface and then diverge. As this occurs, they drag two portions of the original continents apart. This idea has two important consequences. First, mountain building takes place where the convection currents are descending, and second, ocean floor *development* (Holmes's words) takes place on the site of the gap caused when the continents move apart (see Figure 6-1).

What Holmes had identified was no less than a gigantic "con-veyor belt" on a planetary scale.

Was this the first suggestion that continental drift need not in-volve the landmasses ploughing through the basalt of the ocean floor? It was, at least, a quantum leap in our knowledge of the processes that shape the surface of the planet, which is why Holmes is a major player in the story of earth history.

Another figure who joined the continental drift controversy was S. W. Carey (1911–2002), who was a professor at the University of Tasmania. He supported the concept of drifting continents, but his major contribution was his theory of an "expanding earth." When

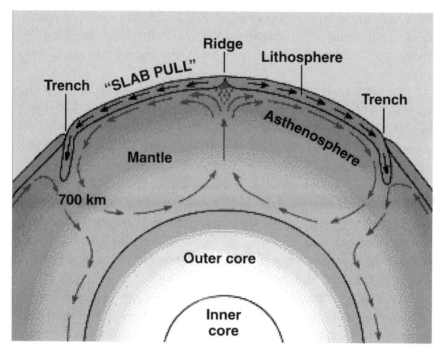

Figure 6-1. The basic structure of the earth. The lithosphere, which includes the crust of the earth together with the uppermost part of the mantle, is directly underlain by the *asthenosphere* which is soft and partially molten. As a result, the lithosphere is effectively detached from the mantle below and can move independently on the asthenosphere.

Hot molten rock in the interior of the earth rises to displace cooler liquid, creating a current and setting up convection cells. The upper limbs of the cells start to rise and eventually touch the base of the lithosphere, where they diverge in opposite directions. As the cells ascend then descend, they drag the continents into movement. In this figure, crust is created at the ridge and destroyed at the trenches, which is part of the process of "seafloor spreading." Credit: United States Geological Survey.

he tried to reconstruct Wegener's supercontinent of Pangaea, Carey found that it was difficult if he assumed that the diameter of the earth had remained fixed. It was, however, considerably easier if the earth had been smaller at the time the supercontinent existed. According to Carey, the earth at the time of Pangaea was only a fraction of the size it is today, and it was completely covered by the supercontinent. Carey then developed an idea that had first been proposed in the 1920s; namely, that the earth had expanded, a process that he attributed to the planet's *heating up*. Carey believed that this growth in size of the earth had caused the breakup of Pangaea, initially splitting it into two pieces, Laurasia and Gondwanaland, and later into more pieces that became the present-day continents. This is the opposite to the theory of the "contracting earth" due to *cooling down*, as proposed by James Dwight Dana in the middle of the nineteenth century. Carey synthesized his ideas in his book *The Expanding Earth*, in which he argues that expansion is the cause of continental drift. The theory of the expanding earth was not, however, popular with most geoscientists, although the arguments surrounding it still rage to this day.

The third strong supporter of Wegener was the South African geologist Alexander du Toit (1878–1948). For many years, he collected evidence in support of the theory of continental drift and brought his findings together in his book *Our Wandering Continents*, published in 1937 and dedicated to Alfred Wegener. In one line of evidence, du Toit found a striking parallel development in a mixed sequence of glacial, marine, and freshwater deposits in both South Africa and South America; later, other workers extended his finding to include India and Antarctica. Du Toit concluded that such parallel developments could only be explained if the two landmasses had once been joined, although he postulated a different initial prior configuration for the continents. Instead of the single supercontinent Pangaea, du Toit argued for two supercontinents: one in the north (Laurasia) and one in the south (Gondwanaland). To support the theory of continental drift, du Toit produced a considerably more detailed body of evidence than Wegener himself had, citing similarities in rock formations, climatic zones, and fossils on both sides of the Atlantic.

Was du Toit's intervention enough to convince the geological world of the validity of continental drift? The answer is no. Not at the time. As Wegener himself wrote in the final 1929 revision of *The Origin of Continents and Oceans*, "It is probable the complete solution of the problem of the forces [driving the movement of the continents] will be a long time coming."

How right he was.

In fact, that proof would only come decades later, as new fields of research were opened up and yielded irrefutable evidence for the movement of continents. But a very important piece of the jigsaw was on the board now, and one thing was for certain: The theory of continental drift changed forever the way we viewed the planet. And, equally important, it opened the floodgates for even more radical ideas that were to lead eventually to the theory of plate tectonics.

7
NEW HORIZONS

THE CHALLENGE OF THE SEA

Before we consider the next of the three great theories that attempt to explain the forces behind the evolution of the surface of the earth—the theory of seafloor spreading—it is necessary to turn back to the seabed and catch up with several disparate lines of marine research.

At the time when Wegener proposed the theory of continental drift, scientists still knew relatively little about the aquatic world that covers two-thirds of the planet's surface, and much of what they did know had come from the *Challenger* expedition. But over the four decades leading up to the publication of the theory of seafloor spreading in the early 1960s, our knowledge of the oceans grew rapidly on all fronts. And, as we shall see, on the road to plate tectonics theory, understanding more about the seafloor was particularly important.

The *Challenger* expedition had made the first scientific ocean survey on a global scale and identified two features that turned out to be key factors in the evolution of the earth's crust: the elevated areas in the mid-ocean regions, and the deep trenches on the ocean margins. In the years following the *Challenger* expedition, many additional details were filled in, and it became possible to identify a number of topographical environments on the seabed. From our point of view, three of them are especially important.

▸ *Continental Margin.* This area includes the continental shelf, the continental rise, and the continental edge. The continental shelf is the seaward extension of the landmasses. Two particular features are associated with the shelf and shelf edge regions. The first are deep, step-sided valleys that cut the shelf at right angles to the land. These submarine canyons are important because they can act as conduits for the transport of material from the continents to the deep-sea. The second are the deep trenches, which, in some areas, lie at the edges of the shelf regions. The trenches are long (up to 4,500 kilometers) and narrow (mostly less than 100 kilometers wide). They are the deepest parts of the oceans and play an important role in deep-sea sedimentation because they can act as traps for material brought across the continental shelf, thus preventing it reaching the deep-sea areas. The trenches are found in all oceanic regions but are concentrated in the Pacific Ocean, where they form a more-or-less continuous ring around the edge of the ocean. See Figure 13-1.

▶ *Ocean-Basin Floor*. This area lies to the seaward side of the continental shelf. Abyssal plains are a principal feature of the ocean basins. These plains are flat regions that extend over thousands of kilometers. They were formed by the action of turbidity currents that transport sedimentary material from the landmasses across the continental shelves and out onto the ocean floor via submarine canyons. Abyssal plains are less well developed in those oceanic areas where sediment-trapping trenches are found. Other features on the ocean-basin floors include abyssal hills and sea mounts (volcanic hills).

▶ *Mid-Ocean Ridge System*. This is a range of mountains that runs through all the major oceans. This major topographical feature of the seabed is important in plate tectonics theory.

Despite a better understanding of the features of the seabed, it was not until the middle of the last century that a topographic map of the global ocean floor became available to scientists. The initial problem in mapping the seabed was the time it took to make soundings—in *Challenger's* time (i.e., the 1870s), it took many hours as the ship held station. What was needed was a way of plotting water depths when a ship was *underway*, because only then could systematic surveys of all the oceans be carried out.

The breakthrough came with the invention of the first sonar

system, the echo sounder, and other technologies developed just prior to, and during, World War II. The principle behind the echo sounder is simple: Water will transmit sound waves, and if a sound pulse is sent from the surface (e.g., via a transducer mounted on the bottom of a ship), it will bounce off the seabed as an echo. The time interval between sending out the pulse and the return of the echo to the transducer is used to measure the depth of water.

Sonar received a boost in WWII when submarine detection was a vital part of the war at sea. Then, in 1954, the precision depth recorder (PDR) was invented, followed later by a variety of other instruments, such as the side-scan sonar, designed to carry out more detailed mapping of the seabed. Side-scan sonar is, as the name implies, a sonar that can look sideways. It sweeps the seafloor from a towfish and the transmitted energy is in the shape of a fan, which sweeps the seafloor to either side for up to 100 meters or more. The echo that bounces back reveals a picture of the ocean floor, but gives no information on depths—unlike the multibeam sonar, which also provides a fan-shaped sweep of the seafloor, but this time the output data is in the form of depths. So oceanographers had a wide range of survey instruments at their disposal, but it was the echo sounder that first revolutionized ocean floor mapping.

MAPPING THE OCEAN FLOOR

The echo sounder offered a relatively fast way of surveying the seabed when a ship was underway. One of the first applications of the echo sounder was on the German *Meteor* expedition (1925–1927). The *Meteor*'s surveys showed that the Mid-Atlantic Ridge, first identified by Matthew Maury, was in fact a rugged mountain range and not the flat, elevated area that Maury had originally thought.

More and more soundings were taken in the following years, but the real breakthrough came after WWII, from a group of scientists at the Lamont Geological Observatory of Columbia University (now the Lamont-Doherty Earth Observatory). The laboratory was established by Maurice Ewing, and during his twenty-five years as director, marine scientists there concentrated on a wide variety of seagoing studies. But it was the production of detailed physiographic maps of the ocean floor, the first comprehensive mapping of its kind ever undertaken, that caught the imagination of the public.

The two leading figures behind construction of the ocean floor maps were Bruce Heezen and Marie Tharp, who began working together on the maps in 1947. Heezen was mainly concerned with gathering data, at sea, on the depth of the ocean bed. The data was taken from more than thirty cruises on the Lamont ship *Vema*, together with data from the *Atlantis*, a ship from the Woods Hole Oceanographic Institution. A wide range of other information, including geological and seismic (earthquake) data, was also used in the construction of the maps. According to Heezen, "Earthquakes helped more than anything to map the ocean." Marie Tharp was the actual cartographer, but did not take part in an oceanographic cruise with Heezen until 1965.

How did Marie Tharp construct the seafloor maps, and how accurate were they? To draw the maps, she used sounding data from which she first plotted seabed profiles, then converted them to a three-dimensional map. Line-drawn contour maps give an indication of topography, but the ocean floor maps went further. The three-dimensional features were colored in by Heinrich Berann, an Austrian artist, using Marie Tharp's original line drawings. The ocean maps reached a wide audience following their publication in *National Geographic* in 1967. The public had their first glimpse of this new world—the approximately 70 percent of the planet's surface that until then had been hidden underwater—and it was revealed in glorious Technicolor.

There is no doubt that the Tharp-Heezen seafloor maps are correct in their general features of the oceanic domain. But, almost by their nature, the maps are broadly stylized, and many of the features were drawn in with a good deal of imagination. This is hardly surprising, considering the vast extent of the oceans and the relatively limited number of soundings, particularly in some remote regions. But perhaps the strongest criticism that can be leveled at them is that they exaggerated the vertical scale, by as much as 20 percent, so that what the public saw in 1967 was really a much distorted picture of the real seabed. Modern techniques, such as multibeam sonar, can give a much more accurate picture of the ocean floor, and indeed large-scale features have been found that were not shown on the Tharp-Heezen maps. But these maps were the first of their kind, and they opened up a whole new subaquatic world. A world of high mountains, vast plains, and deep trenches; a topography as varied as that on any continent.

Heezen and Tharp brought out their ocean floor maps one by one, culminating in their World Ocean map in 1977—the year after Heezen died from a heart attack. What did it look like, this vast area of seabed that was now revealed for the first time? Beyond the boundary of the continental shelves, the depth onto the ocean floor increased sharply, with trenches and lines of volcanoes in some places, leading onto vast, flat abyssal plains. Beyond the plains was the most stunning feature of all, the mid-ocean ridge, an almost continuous mountain range that encircles the globe. It is one of the largest topographic systems on the surface of the earth, extending almost continuously for 60,000 kilometers through the Atlantic, Antarctic, Indian, and Pacific oceans (see Figure 7-1).

The mid-ocean ridge that began to emerge as Tharp was able to input more and more data into her map was a strange, fascinating geological feature. A wide area of gentle slopes led up to a mountain range with a crestal region up to 3,000 meters above sea level, which falls away to gently sloping flank regions. The whole feature was cut by fissures in the surface, offset by fracture zones (or transform faults) and, above all, by an apparent V-shaped cleft or rift (approximately 30–50 kilometers wide) running down the middle—as if somehow the ocean floor was splitting apart.

Marie Tharp had prescient ideas about that V-shaped crack, which she suspected might be a rift similar to the Great Rift Valley on the African continent. She also thought it may even be associated in some way with continental drift.

A rift valley is formed when a block of material subsides, relative to the blocks on either side, between parallel faults as the crust is pulled apart. In cross-section, a rift valley can appear as a steep-sided gorge with a wide floor and a central flat block, much like a keystone that has slipped below the surface of an arch. Rift valleys can be extensive features; for example, the Great Rift Valley runs north to south for some 5,000 kilometers from Southwest Asia to East Africa. This rift valley, like others, is the site of volcanic and geothermal activity (see Chapter 13).

▲

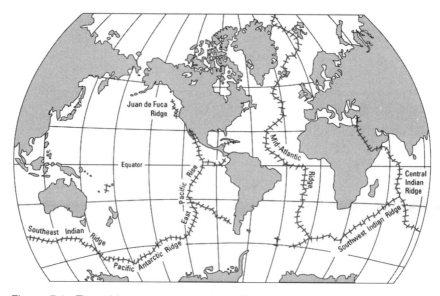

Figure 7-1. The mid-ocean ridge system. The ocean bed maps produced by Bruce Heezen and Marie Tharp revealed a world with a rugged topography as varied as that on the continents. One of the most interesting features was the mid-ocean ridge, a wide area of gentle slopes rising up to a mountain range with deep valleys, steep-sided gorges, and peaks. The mid-ocean ridge is a global feature that occupies 30 percent of the entire ocean floor and is 60,000 kilometers long. The features are cut by fissures and faults that offset the original topography at right angles to the length of the ridge. In particular, the ridges are split by a series of fracture zones, or transform faults, perpendicular to the trend of the ridge. The mid-ocean ridge system is the birthplace of new ocean crust, and plays a pivotal role in the theories of seafloor spreading and plate tectonics. Credit: United States Geological Survey.

Mapping the ocean bed, and identifying the features of seafloor topography, had put another important piece of the plate tectonics jigsaw in place. The next giant step in the story is the introduction of the theory of seafloor spreading, a theory that only saw the light of day after Marie Tharp had hinted that the crack in the North Atlantic mid-ocean ridge might be associated with continental drift. But at the time, her idea was dismissed, and apparently even Bruce Heezen called it "girl talk."

One year later he would change his mind.

8
SEAFLOOR SPREADING

A THEORY EMERGES

The two scientists most associated with the concept of seafloor spreading were H. H. Hess and R. S. Dietz. The idea had been around as a preprint since at least 1960, before it was finally put into the literature in 1962 in the seminal paper written by Harry H. Hess.

Seafloor spreading was more than simply another step along the road to plate tectonics theory, because it marked a major change in attitudes toward moving continents. Up to this point, the single most important argument of the opponents of continental drift theory was the lack of any realistic process that allowed the "lighter" granitic continents to plough through the "denser" basaltic ocean crust. Then seafloor spreading turned everything on its head by proposing that rather than the continents themselves moving, the seafloor *spread apart*. Scientists had at last found the holy grail—the mechanism that could drive Alfred Wegener's theory of continental drift.

The concept was termed *seafloor spreading* by Robert Dietz, and it was an idea that had to some extent been foreseen by Frank Taylor as far back as the first decade of the twentieth century in his proposal that the Mid-Atlantic Ridge was a zone of rifting, and by Marie Tharp, who pushed this view further by identifying the "notch" in the ridge as the location of the rifting. So what exactly did Hess suggest in his famous paper entitled *History of the Ocean Basins*?

A naval officer during World War II, Hess was a professor at

Princeton with a special interest in the history and geology of the ocean basins, and he extended the ideas of convection cells in the mantle put forward by Arthur Holmes by relating them to a mobile seafloor. Hess believed that as these convection cells move, magma escapes from the mantle at the crest of the mid-ocean ridges—the V-shaped notch identified by Tharp—to create new seafloor that spreads apart, pushing at the older seafloor, which "slides" away from the ridges as new material is generated. Dietz's version differed from that of Hess, because Dietz believed that the sliding surface was at the base of the lithosphere, not the crust.

The dynamics of the process require that as new seafloor is *generated*, older seafloor must be *destroyed*, since the ocean floor cannot simply continue to expand on the earth's surface indefinitely. To explain this, Hess proposed that older crust is reabsorbed, or subducted, by sinking into the deep-sea trenches found at the edges of some ocean basins, and continues to disappear under the continents where it undergoes remelting back into the mantle. What Hess was proposing was no less than the recycling of the seafloor: a giant conveyor belt in which new oceanic crust was *created* at the structurally weak zones of the mid-ocean ridge crests and *destroyed* at the ocean edges.

Seafloor spreading answered a question that had long plagued geoscientists—why are the ocean basins so relatively young? The answer is that because the seafloor is being continuously created from the mantle at the ridges, the site of the youngest oceanic crust, and continuously remelted into the mantle at the ocean edges, it cannot be geologically very old. Furthermore, seafloor spreading could explain the existence of two of the most important oceanic features that were initially identified by Matthew Maury and later by the *Challenger* expedition: the mid-ocean ridges (now known to be the site of the generation of new seafloor) and the deep-sea trenches (now known to be the place where the seafloor, with its sediment cover, sinks or subducts under the continents). Perhaps most important of all, in the new theory the mid-ocean ridges were the link between the oceans and the continents in the processes that shape the surface of the earth (see Figure 8-1).

A THEORY COMES OF AGE

Attractive as it must have seemed to many marine geophysicists, Hess's concept of seafloor spreading remained just that—an un-

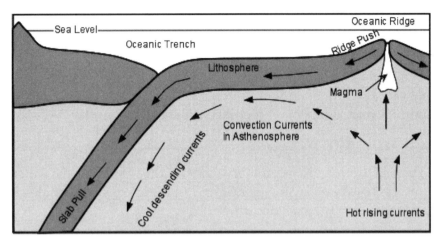

Figure 8-1. The elements of seafloor spreading. Harry Hess proposed that convection cells in the mantle move magma, some of which escapes at the crest of the mid-ocean ridges to create new seafloor. The new floor spreads, pushing at the older seafloor, which slides away from the ridges. As new seafloor is created at the ridges, old seafloor must be destroyed. Thus, a global-scale conveyor belt is operating in which new oceanic crust is continually being created at the mid-ocean ridge crests and continually being destroyed at the ocean edges. The new crust moves by being pushed away at the ridges ("ridge push") and dragged down at the trenches ("slab pull"). Seafloor spreading rates vary from ~1cm/year, for slow spreading ridges, to 8cm/year for fast spreading ridges. Credit: S. Nelson, "Global Tectonics."

tested hypothesis. But, like many scientific theories, once it was out in the open, supporting evidence began to pour in from a variety of sources. This evidence should be seen against the background of what was known about the oceanic crust.

Seismic research has added greatly to our knowledge of both the thickness and structure of the oceanic crust. One of the first findings to emerge was a confirmation that the crust under the oceans is very different from that under the continents, especially with respect to seismic layering. Under the continents the crust appears to be variable in structure, but generally the top layer, between 10–20 kilometers in thickness, is composed of rocks having seismic velocities characteristic of granites. In contrast, the oceanic crust consists of at least three distinct layers, one lying on top of another. The petrology of these layers, which is inferred from seismic and drilling data, is extremely complex. For simplicity, however, a number of the layers in the oceanic crust can be identified.

Layer 1 is composed of either unconsolidated or consolidated sediment, or both.

Layer 2, the oceanic basement on which the sediment layer is deposited, consists mainly of volcanic basalt lavas, overlying metamorphic and sedimentary rocks (consolidated sediment).

Layer 2A consists of fractured pillow lavas formed when lava is extruded from a fissure underwater. As this happens, the outer crust is cooled quickly by quenching, but the interior of the pillows cool more slowly from conduction.

Layer 2B also consists of basaltic pillow lavas, but now the spaces between them are occupied by clays and products of metamorphism.

Layer 2C is composed of the intrusive equivalent of basalt in feeder dike swarms. The swarms are emplaced at the same time during a single intrusive event, and the feeder dikes move through fractures, about one meter wide, and transport magma intrusively upwards to the seafloor to produce lavas.

Layer 3 is believed to be made up of gabbros—dark, course-grained igneous rocks chemically equivalent to basalts, but plutonic in character. They are formed beneath the earth's crust when magma of the right composition cools and crystallizes. (The Moho lies below layer 3.)

The evidence to support seafloor spreading came from several quarters, including the mid-ocean ridge system, heat flow on the ocean bed, and gravity measurements.

The Mid-Ocean Ridge System. One line of evidence came from the seabed topographical maps that Bruce Heezen and Marie Tharp continued to produce. The most important single finding to emerge from these maps was the confirmation that the mid-ocean ridge, and its central notch, was a global feature; thus providing evidence that seafloor spreading worked on a global scale.

Heat Flow in the Mantle. Convection cannot take place without a source of heat, and one geophysical tool that added to the verification of seafloor spreading was the measurement of heat flow on the ocean bed. Essentially, heat flow is the movement of heat. It can be defined as "the product of the increase in temperature with depth (gradient) times the thermal conductivity." In the oceans, the only cold surface through which heat can be lost is upward through the seabed, so heat flow may be regarded as the amount of heat emanating from the ocean floor.

Questions surrounding the earth's "heat budget" had puzzled scientists for a long time. To begin with, it had been assumed that the earth had cooled *progressively* from its hot primordial state. Sir Isaac Newton, and later Lord Kelvin, used the concept of a cooling planet to estimate the age of the earth, and it wasn't until the realization that radioactive decay produces heat that the idea of the cooling world was challenged.

Now it is known that the interior of the earth has two principal sources of heat. First, there is heat from the radioactive decay of elements, mainly uranium, thorium, and potassium; this heat migrates upward, toward the surface of the earth. Second, there is the residual heat left behind when the planet was formed. The temperature of the earth increases with depth, which gives rise to a geothermal gradient, and the geothermal flux is the influx of heat to the surface from below; that is, from the earth's internal heat.

Under the "thicker" continental crust, the heat flow is made up roughly of 80 percent radioactive heat and 20 percent residual heat. But under the "thinner" oceanic crust, radioactive heat contributes only 10 percent of the total heat; the rest being made up mainly from other sources, such as convection currents that can transfer heat upward in certain areas.

Convection is the transfer of heat by the movement of a mass, or substance, from one place to another. The idea that convection cells in the mantle were the engine that generated seafloor spreading, by releasing upwelled molten rock at the mid-ocean ridge crests, had profound heat flow implications. Heat flow is determined by measuring temperature change with depth, with the gradient recorded being a function of the heat flow—provided, of course, that the measurements are made in the same material. The conditions controlling heat flow are extremely complex, but some general patterns can be identified on a global scale:

▸ There is a decreasing heat flow with increasing age of the seafloor.

▸ Variations in heat flow occur within the oceans, and a clear pattern began to emerge as more and more heat flow data came in. The first measurements, made by Sir Edward Bullard in 1954, indicated that high heat flow occurred over the mid-ocean ridges, or as they became known later, the centers of seafloor spreading. Over the following years, as more and more measurements were made from probes pushed into the sediment cover, it became apparent that, despite the very large scatter in the data, the ridge crests are indeed characterized by high heat flow values that are six to eight times the average for the ocean. In contrast to the ridge areas, relatively low heat flow values have been reported for the deep-sea trenches. Overall, the pattern of heat flow in the oceanic crust reflects the production of new crust at the ridge centers and its cooling on the ridge flanks as it moves away during seafloor spreading. As the hot magma escapes at the ridge crests, it provides a mechanism for much of the heat loss from the earth's interior.

If seafloor spreading did take place, as was assumed, then the heat loss from the seafloor would be highest around the spreading ridges where hot magma was brought up from below. This was indeed the case, adding further weighty evidence in support of seafloor spreading. But then a very interesting phenomenon made its appearance. Evidence began to emerge that in some mid-ocean ridge locations the heat flow was much lower than expected. So where was the "missing heat"? The fact that heat was missing could have thrown a wrench into the acceptance of the theory of seafloor spreading, but in fact it led to the suggestion that some process other than conduction might be cooling the crust in the mid-ocean ridge areas—an extremely important conclusion that will be discussed later when hydrothermal vents are described.

Gravity Measurements. Gravity anomalies, which result from deviations in excess of those that may be considered normal, can be positive or negative, and they have been used to interpret structure in the material beneath the crust. Gravity surveys revealed that most of the ocean floor is in isostatic equilibrium, with one major exception: In the deep-sea trenches, there are strong negative anomalies that have been interpreted as resulting from a down-buckling, or subduction, of parts of the oceanic crust; thus confirming one of the principal pillars of seafloor spreading—the subduction of the oceanic crust under the continents.

THE PALEOMAGNETIC KEY

Although heat flow, together with seismic and gravity measurements, provided evidence to support seafloor spreading, there is little doubt that the strongest evidence of all came from paleomagnetic studies.

The earth has a magnetic field, and when rocks formed at high temperature cool down they reach the Curie point, below which the magnetic (iron-rich) minerals in them become magnetized. During this process, their polarity (the north-south orientation) is assigned to that of the earth, giving rise to a *normal* polarity. Then as the rock cools down this polarity is frozen in a permanent alignment in the solidified magnetic minerals and provides a "fossil" record of the earth's magnetic field at the time the minerals were formed—paleomagnetism. The ocean crust is particularly appropriate for magnetic studies because the principal rock, basalt, contains the mineral magnetite.

As geophysicists began to study the seafloor with magnetometer surveys, it became apparent that a magnetic paradox exists in which some minerals have a *reversed* polarity with respect to present conditions. The reason was found by studying age-dated rocks on land. From these studies it became clear that the earth's polarity had undergone a series of switches in the past, during which the north and south magnetic poles have flipped over and changed places.

In the story of seafloor spreading, the most important application of magnetic techniques was the measurement of polarities around the mid-ocean ridges. The initial studies were made by two scientists, Ronald Mason and Arthur Raff, from the Scripps Institute of Oceanography. They towed magnetometers over the seafloor off the western coast of the United States and reported their findings in 1961. They found that there was a good deal of structure in the magnetic data, and that patterns could be traced for hundreds of kilometers on the ocean floor. This information was used to interpret the tectonic history of the region, and when the data was plotted on a magnetic map it revealed a series of magnetic strips—one strip for minerals with north pointing to the current earth north (normal polarity) and one for minerals with north pointing to the current earth south (reversed polarity). Close examination of these magnetic strip patterns showed that many of them were offset, which was interpreted as evidence that the seafloor had been cut by faults into major fracture, or dislocation, zones.

At this stage, although magnetic studies were providing interesting information on crustal evolution, it was not clear how they might be related to the structure of the ocean floor. Then Lawrence Morley, a Canadian paleomagnetist, came up with the idea that the magnetic anomalies were the result of the interaction between seafloor spreading and magnetic reversals in the rocks generated as new crust was formed. However, Morley was not taken seriously enough, and the full implications of the magnetic anomalies were not revealed until Fred Vine and Drummond Matthews published their now-famous 1963 *Nature* paper.

If Hess and Dietz provided the theory behind seafloor spreading, then Vine and Matthews provided incontrovertible proof that it worked. Vine and Matthews were marine geophysicists working at Cambridge University, and their 1963 paper provides an excellent example of a major scientific breakthrough that was, in fact, based on a somewhat-limited data set. The stroke of genius shown by the two scientists was their ability to make the quantum leap and extrapolate the limited data of a local phenomenon to a global-scale process.

Vine and Matthews found "parallel zebra strips of normal and reversed magnetism" in the region of the Carlsberg Ridge, part of the mid-ocean ridge system in the Indian Ocean. But the great importance of their work was that the polarity of the strips was symmetrically distributed on either side of the ridge—in other words, a strip of normal magnetism would be followed by one of reversed magnetism, then by another normal magnetism, and so on, on each side of the ridge crest. The explanation proposed to account for these alternating magnetic stripes in their striking linear patterns went a long way toward confirming the existence of seafloor spreading.

Vine and Matthews proposed that as new oceanic crust, in the form of basaltic lava, wells up and escapes from the crestal regions of the ridges to form new seafloor, it solidifies and records the ambient magnetic field. This process forms a strip of rock, showing ambient magnetism, along the crest of a mid-ocean ridge. The width of this strip of rock with the ambient polarity will depend on two factors: the length of time before the next "flip" of the poles, and the rate at which the seafloor spreads apart. Then as newer basalt wells up, it pushes the older rock away from the ridges, and when the earth's magnetic polarity undergoes intermittent reverses, they are recorded in the next strip of rock formed. In this way, a symmetrical pattern of magnetic polarities is frozen in the rocks as they move away in opposite directions on either side of the mid-ocean ridge (see Figure 8-2). Although Vine and Matthews

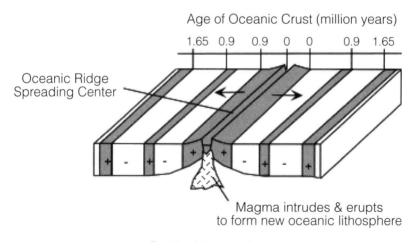

Figure 8-2. The proof of seafloor spreading. Vine and Matthews showed that normal positive (dark strips) and reversed negative (white strips) magnetic patterns indicated the emplacement of newly formed basalts on either side of the mid-ocean ridges in strip-like arrangements. Credit: S. Nelson, "Global Tectonics."

were unable to demonstrate the absolute symmetry of the strips, the theory was in place and came to prove the validity of seafloor spreading.

Many magnetic studies of seafloor rocks followed the pioneering work of Vine and Matthews, and it became apparent that very large areas of the ocean basins are characterized by linear magnetic anomalies associated with seafloor spreading— confirmation that it was a global phenomenon.

Scientists knew that the earth's magnetic field changed polarity intermittently through time, but at first it was impossible to date the reversals. Then, in 1966, the Jaramillo Event was identified. This was a reversal in the earth's magnetic field that occurred 900,000 years ago, and it could be identified in many places. From this finding, geologists working for the United States Geological Survey (USGS) at Menlo Park dated magnetic orientations in volcanic rocks using radioactive techniques to establish a timescale (the *geomagnetic polarity timescale*) for magnetic reversals. It is now thought that there have been around 170 such reversals over the last 76 million years, and that the present magnetic orientation has been in place for around 600,000 years. Magnetic polarity measurements therefore offered geophysicists a tool to look back into the history of the earth's formation, and later the magnetic reversal timescale provided the key to unscramble the history of seafloor spreading.

HOW FAST DOES IT WORK?

As seafloor spreading takes place and new crust is continuously pushed away from the ridges, it follows that the ocean floor furthest from the ridges is the oldest. At first, little was known of the rate at which the seafloor was spreading apart. But eventually, the timescale for magnetic reversals was devised for the last 4 million years and allowed the width of a magnetized stripe to be related to the rate of seafloor spreading. Early studies showed that the spreading rate was on the order of one to ten centimeters per year.

To calculate the rate of seafloor spreading, it is necessary to know how far a portion of the crust has moved from the point at which it was generated. This is not always easy because of the offsets caused by faulting at the ridges. Another problem is that away from the ridges, seafloor spreading does not take place in a simple lateral way, but instead may involve rotation about an axis.

Eventually, it became apparent that spreading rates varied from one oceanic locality to another; some regions reflect "slow" rates and some "fast" rates. The factors controlling the rates of spreading are complicated, but steep ridges appear to have slower spreading rates (e.g., one or two centimeters a year for the Mid-Atlantic Ridge) than a less steep system (e.g., five to eight centimeters a year for the Pacific).

Seafloor spreading theory contributed what turned out to be the pivotal piece of evidence to the plate tectonics jigsaw, and incontrovertible evidence was now piling up to support the plate tectonics theory as the heir to seafloor spreading continental drift. However, there were still vast gaps in our knowledge of the seafloor to be filled in as Tharp and Heezen continued to refine their maps. The seabed they portrayed was covered with a blanket of sediment, but in fact very little was known of the nature of the blanket below the first few tens of meters.

But this was to change as new research programs continued to probe the secrets of the oceans.

9
THE PLOT THICKENS

DEEP-SEA SEDIMENTATION

The seabed is composed largely of basalt, which in most, but not all, places has a cover of sediment.

The oldest deep-sea sediments are Cretaceous in age—that is, they are no more than 140 million years old, while the earth itself has been in existence for around 4.55 billion years (see Chapter 2). Deep-sea sediments therefore represent no more than 3 percent of geological time, which raises the question: Why aren't there older oceanic sediments? We will return to this question later.

The oceanic sediment layer is comparatively thin, with a global ocean average of 500 kilometers. There are variations in sediment thickness in individual oceans that appear to follow reasonably well-defined patterns. For example, in the Atlantic Ocean the thickest sediments are found around the shelves and edges of the ocean basins where they can be over 1,000 kilometers thick, and the thinnest in the central mid-ocean ridge areas. The American geochemist Karl Turekian has estimated that at the rate we know the sediments have been accumulating, it would only have taken about 1,000 years for the sediments at the Mid-Atlantic Ridge to have accumulated and no more than 50 million years for the thickest Atlantic deposits to have formed.

So, until the theory of seafloor spreading emerged, two important questions puzzled geologists: Why was the oceanic sediment cover so thin in mid-ocean areas, and why was the oldest sediment so young compared to the age of the whole earth?

Prior to the *Challenger* expedition (1872–1876), little was known about the sediment covering the seabed, or its thickness, and no systematic oceanwide study of these deep-sea deposits had been carried out. Such a study was started by John Murray during the expedition, and completed later using both material in the *Challenger* collection and samples collected by other ships. The results of the study were published by John Murray and A. Renard in 1891 in the *Challenger Report on Deep-Sea Deposits*. This was a landmark volume that attempted to classify the main types of deep-sea sediment and map their distributions on the seafloor. It was an enormous task, and laid many of the foundations of marine geology for future generations.

Although the number of samples available to them was still relatively small considering the vast size of the oceans, Murray and Renard were able to establish two major trends in the distribution of the deep-sea deposits:

▸ Large areas of the seabed at water depths less than 2,500 fathoms (4,557 meters) are covered with a calcareous sediment termed *globigerina ooze,* after the predominant shell-forming organism.

▸ Other areas of the seabed, at water depths greater than 2,500 fathoms are covered by *red clay.*

An important concept was therefore beginning to emerge—that the distribution of deep-sea deposits was governed by processes that operated on a *global-ocean* scale. As marine geology advanced, more information became available on the nature of the sediment layer that covers the seabed and on the distribution of the major types of marine sediments. As this knowledge continued to grow, the trends first hinted at by Murray and Renard were confirmed.

Marine sediments can be divided into two main types: near-shore and deep-sea. Near-shore sediments are laid down on the continental shelf regions under a wide variety of sedimentary environments (such as deltas, estuaries, bays, lagoons) and are strongly influenced by local landmasses. *Deep-sea* sediments, which cover more than 50 percent of the surface of the globe, are deposited beyond the continental shelf in depths of water that are usually greater than 500 meters. Because of factors such as remoteness from the sources of sedimentary material on land, great depth of water, and a distinctive oceanic biomass, the environment in which deep-sea sediments are laid down is unique on the planet.

In general terms, deep-sea sediments consist primarily of two types of material:

1. The debris of shells of tiny planktonic organisms, such as foraminifera (calcium carbonate shells) and diatoms (silica shells). Sediments with more than 30 percent of these shell remains are termed *oozes*.

2. Fine-grained inorganic material derived from the land-masses and brought to the oceans via river, atmospheric, and glacial transport. The land-derived material is mainly inorganic and has two important characteristics: It has a very small particle size, usually less than 2 micrometers, and it is deposited at a relatively very slow rate—usually a few millimeters per thousand years. Deep-sea sediments composed mainly of inorganic components and having less than 30 percent shell remains are termed *clays*. In addition, the clays are subdivided into hemipelagic and pelagic types on the basis of the route by which their components were transported to the seabed.

There are several mechanisms by which sediment material can be brought to, and distributed within, the oceanic environment. The two most important are horizontal transport off the shelves and along the seabed by gravity currents, and vertical transport down the water column.

Ocean bed gravity currents move material that was originally deposited on the continental regions to the deep-sea by processes such as slides, slumps, and gravity flows. The most important of the gravity flows are *turbidity currents,* which are short-lived, high-velocity, density currents that transport vast quantities of sediment across the shelves to the floor of the deep-sea, often via submarine canyons. As this happens, the sediments spill out and form the vast abyssal plains that fringe the continental shelf areas, particularly in those regions where there are few trenches to trap the material before it can spread. Sediments deposited by turbidity currents, such as those of the abyssal plains, are termed *hemipelagic* deep-sea sediments.

Down-column transport, which is also gravity-driven, carries material vertically down the water column from the surface ocean to the seabed. This form of transport carries both land-derived material and planktonic shell remains down the water column. Sediments deposited from material transported down the water column are termed *pelagic* deep-sea sediments.

Gravity current transport has its major effect on the regions that fringe the continents, and it is here where the abyssal plains are formed. Vertical down-column transport operates over all the oceans, but has its greatest effect in the more remote areas— including those around the mid-ocean ridge system.

So, the two main deep-sea sediments are oozes, consisting of calcareous and siliceous shell remains, and clays, consisting mainly of land-derived inorganic material. In addition, glacial marine sediments, rich in glacial debris, are found in some oceanic regions. These major sediment types are deposited in a clear global-ocean pattern, and a number of large-scale trends can be identified:

▶ The calcareous oozes cover large tracts of the open-ocean deep-sea bed, at water depths less than 3–4 kilometers, especially around the mid-ocean ridges. The reason for this depth constraint is that carbonate shells are more soluble as the pressure, and therefore the depth, of the water column increases. There is a boundary termed the calcium *carbonate compensation depth* (CCD), above which carbonate shells will accumulate in the sediments and below which they will dissolve.

▶ The siliceous oozes form a ring around the high latitude ocean margins, especially in the Antarctic and North Pacific and in a band in the equatorial Pacific.

▶ The deep-sea clays cover large areas of the ocean floor at water depths below the CCD. Hemipelagic clays stretch away from the edges of the oceans, and pelagic clays are found in remote mid-ocean regions.

▶ Glacial marine sediments are largely confined to a band surrounding Antarctica, and to the high latitude North Atlantic.

PROBING THE SEDIMENT BLANKET

Originally, sediment samples from the seabed were obtained by coating the lead weights that were used to take soundings with a substance such as tallow and bringing to the surface anything that attached itself to it. Subsequently, techniques improved and corers were designed to take samples of the sediment column.

There are several types of corers. A box corer retrieves a large undisturbed rectangular sediment sample. A gravity corer is essentially a tube with a heavy weight attached; the weight drives the tube into the soft sediment layer, and a catcher prevents the material escaping as the corer is brought back to the surface. At first core samples were relatively short, but things changed with the introduction of the piston corer, which has an internal piston that is triggered when the apparatus hits the seabed. The piston helps to preserve the sediment core in an undisturbed form and allows the oceanic sediment layer to be sampled to a depth of 20–30 meters.

Being able to take long cores opened up a new world for marine scientists. These cores contain records of ocean history and provide information on the structural development of the ocean basins, water circulation patterns, and climate change. Later, seabed sampling was given a massive boost when drilling overtook core collection techniques and it became possible to retrieve samples of the whole length of the deep-sea sediment column.

When the theory of seafloor spreading saw the light of day in 1962, however, our knowledge of the lower sediment column and the seabed underlying it was sparse. To fill the gap would be costly, and would eventually require an effort on an international scale. But when it happened, the rewards were great and, among many other advances in our knowledge of the seabed, strong evidence emerged to support the theory.

The Mohole Project (1958–1966) was devised to drill through the seafloor into the earth's mantle as far as the Moho—the boundary at the base of the crust (see Chapter 2). In phase I, which ran in the spring of 1961, five holes were drilled under the seafloor off the coast of Mexico, and the deepest showed sediment of Miocene age (approximately 5 million to 23 million years ago) lying on basalt. However, phase II of the project was abandoned in 1966 due, among other things, to cost overrun. But deep-hole drilling at sea was a proven technique by now, and other projects followed.

The Joint Oceanographic Institutions for Deep Earth Sampling

(JOIDES) was established in 1963 by the U.S. National Science Foundation. JOIDES set up a program for deep ocean sediment coring among a group of U.S. institutions that included Scripps Institute of Oceanography, Woods Hole Oceanographic Institution, and the Lamont-Doherty Earth Observatory. JOIDES was an ambitious enterprise conceived with the aim of investigating the evolution of the ocean basins by obtaining drill cores of the entire sediment layer, together with the top of the underlying basalt basement of the oceanic crust.

The first major program set up by JOIDES was the Deep Sea Drilling Project (DSDP). To carry out the surveys, the *Glomar Challenger* was launched in 1968. The ship used "dynamic positioning" over a seabed sonar sound source to hold station and could operate in depths of water up to 6,000 meters. From 1968 until she docked for the last time in 1983, *Glomar Challenger* logged more than 600,000 kilometers, drilled more than a thousand holes at 624 sites, and recovered a grand total of almost 100 kilometers of sediment core from all the major oceans—the exception being the Arctic.

Despite serious problems, such as incomplete recovery of the sediment record and disturbance of the samples actually retrieved, the cores collected by the *Glomar Challenger* have provided invaluable evidence of how the oceanic sediment blanket was formed. Marine geologists now had samples from all depths in the oceanic sediment column and, for the first time, they could begin to work out a detailed history of the ocean floor. Even so, there were others who still doubted the validity of seafloor spreading and instead continued to support James Dwight Dana's view from the previous century that the continents and oceans were fixed.

So the question at the time was, Would the new DSDP history of oceanic sedimentation support the requirements of seafloor spreading?

The answer was to be an emphatic yes.

THE HISTORY OF THE SEABED

Three particularly important discoveries relating to seafloor spreading were to emerge from the DSDP.

The first was to confirm how youthful the ocean floor actually is. It's no older than 200 million years, compared to an age of 4.5 billion years for the whole earth.

The second discovery was related to the thickness and distribution of deposits within the oceanic sediment column. Although the factors that control the thickness of the sediment layer at any one point in the oceans are complex, the general trend found in the data from the DSDP confirmed that the sediment cover was almost absent directly over the mid-ocean ridges and increased in thickness away from the ridge flank. This is entirely consistent with the basic hypothesis of seafloor spreading, which states that the crust at the ridges is young and has therefore only had time to accumulate a thin sediment cover, whereas further away from the ridges the crust is older and has had time to accumulate a thicker cover.

The type of sediment deposited on the seafloor is strongly dependent on the depth of water, with the CCD providing a cutoff between the distribution of deep-sea carbonates and deep-sea clays; above the CCD carbonate sediments predominate, whereas below this boundary carbonate sediment should not be preserved and clays are predominant. The problem was, however, that the DSDP data revealed that carbonates are found in the sediment column *below* the CCD where they should not have been preserved if the carbonate shells had fallen down the water column to those depths. One possible explanation is that the depth of the CCD has fluctuated with time. However, this does not appear likely because the age of the boundary between carbonate and clays is not the same everywhere. Assuming, therefore, that the depth of the CCD has not undergone major changes with time, the most rational explanation for the distribution of the sediments is that it is a function of seafloor spreading.

In this way, if it is assumed, as it indeed is in the Atlantic, that the mid-ocean ridge runs down the center of the ocean, a general model based on seafloor spreading can be set up to describe the sedimentation regime that occurs in the oceans. In this simple model, the first assumption is that the initial sediment, which is laid down as the new ocean spreads away from the ridges in strips, will be a carbonate ooze formed as shell remains sink down the water column and are preserved as oozes in the relatively shallow water depths around the mid-ocean ridge system. As the new oceanic crust is transported further away from the ridge seafloor spreading centers, it is carried below the CCD, and any carbonate shells sinking down the water column at that stage will be dissolved. Thus, the surface sediments will be dominated by clays, which lie on top of the carbonate sediments transported from the ridge areas. In this way, carbonate sediments can be found below the CCD. In some places around the edges of the oceans, siliceous-rich sediment is deposited.

But there was a sting in the tail, because, in an unexpected twist, it turned out that carbonate oozes were *not* the first sediment to be deposited on the basaltic seafloor. Rather, the honor went to a thin red-brown layer at the base of the sediment column that appeared to have been the first deposit laid down there (see Figure 9-1). As it subsequently turned out, this red-brown layer held one of the great secrets of seafloor spreading, which we will return to later, in Chapter 12.

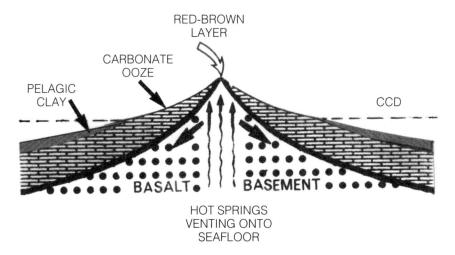

Figure 9-1. A model for sediment accumulation at the spreading ridges. Sedimentation at the spreading ridges presented two enigmas. First, the initial sediment deposited on the new ocean floor, which is above the CCD, should have been an ooze composed of carbonate shells that have fallen down the water column. In fact, it turned out to be a red-brown deposit. Second, carbonate oozes should not have been preserved at depths below the CCD; but in fact the oozes were found below this boundary in off-ridge positions. It is now known that the red-brown layer is a hydrothermal sediment, precipitated from hot springs debouching on the seabed at the spreading ridges. The sediment deposited on this red-brown layer was, as expected, a carbonate ooze, preserved at the relatively shallow water depths that are above the CCD. As new crust is formed and the seafloor spreads apart, however, the carbonate ooze is carried with the migrating ocean floor into deeper water away from the ridges where it is covered, and protected, by a blanket of clay. Credit: Modified from a number of original sources.

The third of the discoveries emerged on DSDP Leg 3, and it stands out as one of the classic groundbreaking studies in planetary evolution. During Leg 3, the *Glomar Challenger* sailed across the equatorial Atlantic from Dakar to Rio de Janeiro. Magnetic anomaly measurements had already shown that the age of the seafloor could be predicted using the magnetic reversal timescale. One crucial test of seafloor spreading would therefore be to compare the magnetic age of the seafloor *rocks* with the paleontological age of the *sediments* resting on them at the base of the sediment column. The data obtained from Leg 3 showed an almost perfect match between the two ages. By making this link between the basaltic seabed and the sediments deposited on it, and by demonstrating that the rocks of the basement were younger toward the ridge centers, seafloor spreading theory had come of age.

Seafloor spreading was able to answer the two important questions related to deep-sea sedimentation:

1. *Why is the sediment cover so thin in mid-ocean areas?* Because the crust at the ridges is the youngest in the oceans and has only had time to accumulate a thin sediment cover.

2. *Why is the oldest sediment relatively young compared to the age of the earth?* Because the ocean basins themselves are relatively young planetary features.

Data poured in from many sources to support the theory of seafloor spreading. In particular, the DSDP provided a necessary piece of the plate tectonics jigsaw by providing a window on the whole thickness of the oceanic sediment column and showing that the data obtained from the deposits found there was fully compatible with the notion of seafloor spreading. But what do we know of the dynamics of the process? To answer this question it is necessary to look more closely at the environments of the mid-ocean ridges (the *birthplace* of the oceanic crust) and the deep-sea trenches (the location of the *destruction* of the oceanic crust) and, in particular, at the mantle lying below them.

10
THE CYCLE OF BIRTH, LIFE, AND DESTRUCTION

THE DYNAMICS OF SEAFLOOR SPREADING

A number of mechanisms have been proposed to explain how chunks of the earth's crust can be moved around. It is generally agreed, however, that large-scale movements, such as those involved in seafloor spreading, can only be adequately explained by the viscous drag resulting from convection currents in the mantle—an idea that had been strongly supported by Arthur Holmes (see Chapter 2).

Viscous drag is therefore a fundamental requirement for seafloor spreading, although the earlier idea that the ridge axes are situated over the rising limbs of large convection cells in the mantle, and that trenches occur over descending limbs, has been largely discarded as being too simplistic. Still, it is convenient for now for us to assume that mantle convection happens, and that as a consequence, two forces shape the ocean floors.

These forces are the "push" from the birth of new crust that thrusts itself up at the mid-ocean ridges, and the "pull" from the destruction of old crust as it sinks below the continents at the trenches. In this way, the floor of the ocean can be divided into three topographic sectors: the mid-ocean ridge, which may be regarded as the place of *birth* of the oceanic crust; the ocean floor, where the crust lives most of its *life*; and the trenches, where the crust *dies*. The history of the ocean crust can be traced through this cycle of birth, life, and death.

123

Birth: The Furnace of Creation. The mid-ocean ridge, which occupies around 30 percent of the entire ocean floor, is the largest mountain chain and the most active system of volcanoes in the solar system. At 60,000 kilometers long, but only 530 kilometers wide, it is one of the most mysterious realms in the entire ocean world. If it could be viewed directly, it would emerge as a rugged topography of deep valleys, steep-sided gorges, and peaks as high as 3.5 kilometers in the Atlantic Ocean. The ridge features are all cut through by fissures and faults that offset the original topography at right angles to its length. In particular, the ridges are split by a series of great fracture zones, or transform faults, mostly perpendicular to the trend of the ridge, which allow bits of the system to slip past one another horizontally. And all the time, the crest is cut by the V-shaped cleft of a valley that is as much as a mile deep and around 20 miles wide. It was this cleft that fascinated Marie Tharp as she drew her seafloor maps.

These are the features of the mid-ocean ridge in the Atlantic, as mapped by Heezen and Tharp. In reality, however, the Atlantic mid-ocean ridge is not typical of the global ridge system. Rather, it is one end of a spectrum, the other being the mid-ocean ridge in the Pacific. Here, the "mountains" are not as high as in the Atlantic, the slopes are gentler, and the features much less rugged. In particular, the central V-shaped cleft at the center is not as well developed as it is in the Atlantic system.

These differences at the two ends of the mid-ocean ridge system reflect the histories of the oceans in which they are found. In geological terms, the Atlantic is younger than the Pacific, and there are a number of important differences between the two oceans, in addition to the topography of their mid-ocean ridges:

▸ The rate of seafloor spreading is faster in the Pacific (roughly 5–8 centimeters per year) than in the Atlantic (roughly 1–2 centimeters per year).

▸ The Pacific is edged with a ring of trenches, whereas there are considerably fewer trenches in the Atlantic.

▸ In the Pacific Ocean, the trenches are consuming crust at a faster rate than it is being created at the ridge, with the result that the Pacific is shrinking.

In terms of the development of its mid-ocean ridge system, the Indian Ocean may be considered to be intermediate between the Atlantic and the Pacific end-members. An interesting feature in the Indian mid-ocean ridge is the presence of a *triple point* that develops when three spreading centers intersect.

New crust is generated in the "furnace of creation" along the ridge crests within a zone of just a few (five or less) kilometers around the spreading center. Furthermore, magnetic anomalies indicate that along much of the global mid-ocean ridge system, the production of new crust is symmetrical around the ridge crest. Asymmetrical spreading has, however, been reported in the Galapagos region in the Pacific Ocean.

The axial region of the mid-ocean ridge is the most exciting part, and two factors can lead to variations in the topography of this area.

The first is the rate of seafloor spreading. At slow-spreading rates (1–4 centimeters per year), a deep rift valley (1–3 kilometers deep) is developed at the ridge axis, whereas at faster-spreading rates (greater than 8 centimeters per year), the axis is characterized by an elevation of the seafloor of several hundred meters, called an axial high. At intermediate spreading rates (4–8 centimeters per year), the ridge crest may have either a rift valley or an axial high. Overall, therefore, the major topography of the mid-ocean ridge system is governed by the seafloor spreading rate. Slow-spreading ridges are narrower (2,000 kilometers), taller, and smoother, whereas fast-spreading ridges are wider (4,000 kilometers), lower, and rougher.

The second factor affecting the morphology of the axial ridge regions is the rate at which magma is supplied from below. For example, a very fast rate of supply of magma can produce an axial high even when the seafloor spreading rate is low.

Finally, it must be stressed that the global mid-ocean ridge system is a discontinuous structure that is offset at right angles to its length by transform faults.

To understand the processes involved in the dynamics of seafloor spreading, we can start with the fact that the ocean floor is formed into a layered "gabbro-dike, swarm-pillow lava" sequence (the layered ocean crust, as described in Chapter 8) from magma stored under the mid-ocean ridge system.

In the semimolten material of the mantle, a series of convection cells have been established (see previous chapters for Figure 6-1 and Figure 8-1). These cells transport the semimolten mantle around the interior of the earth in slow, ponderous movements that bring material to where the mid-ocean ridge system has been formed at the confluence of two ascending cell limbs. The upper mantle is composed of peridotite, which partially melts to form basaltic magma that separates from the unmelted mantle. At the ridge crests, this magma rises up in response to convection currents and accumulates in magma chambers. These magma chambers lie at a depth of approximately 2 kilometers below the ridges, but an actual eruption in the "furnace of creation" emerges from a cupola of molten lava at the top of a chamber, and takes place via dikes to form pillow lavas, which are quickly quenched on the seafloor. As more basalt is formed, the seafloor is pushed apart to make room for it. In this way, seafloor spreading takes place and the "gabbro-dike, swarm-pillow lava" layered sequence of the ocean floor is formed.

The relationship between the magma chambers and volcanic activity is complex. In the ocean ridge system, the presence of the magma chambers appears to be episodic. Thus, the material in them heats, cools, then solidifies to yield frozen magma chambers that can be reheated by another volcanic event, so sustaining repeated volcanic activity for millions of years as the seafloor spreads apart on either side of a ridge.

Life: A Journey Across the Bottom of the Abyss. Following its birth at the spreading centers, the new oceanic crust begins a journey that carries it across an area that becomes a new ocean basin. As more oceanic crust is made, the ocean basin expands until the earlier-formed floor has crossed vast spaces.

What happens to the crust in the perpetual darkness at the bottom of the sea? Undersea volcanic activity can occur, and the next major evolutionary step is that it acquires layer 3, the sediment cover, that completes the layered structure of the oceanic crust. When the ocean floor sinks beneath the continents during subduction, it carries the sediment layer with it. This is important because the sediments contain volatiles that cause intense volcanic and earthquake activity at the sites where the crust is destroyed. The ocean bottom also contains carbonate sediments, trapped there beneath the CCD (as described in Chapter 9) as the crust is moved into deeper water by seafloor spreading. We will be returning to this subducted carbonate sediment in Chapter 16, when we look at its role in the global atmospheric CO_2 budget.

Death: The Tomb of Destruction. To accommodate new crust that is formed at the spreading centers, old crust is destroyed at the trenches in what Harry Hess described as a colossal "ridge creation/trench destruction" conveyor belt.

It was perhaps something of a surprise when the early oceanographers, like those on the *Challenger* expedition in the nineteenth century, found that the deepest parts of the oceans lay not in the center of the basins, but in trenches at the margins. Later, when the topography of the seabed became better understood, it was apparent that these trenches are widespread—particularly in the Pacific, where they form an almost unbroken ring around the edge of the ocean. The trenches are elongated, narrow, deep, V-shaped features of the ocean floor that can be up to thousands of kilometers long, although they can be broken up into smaller units separated by sills. The trenches, which are usually less than 8 kilometers wide, are the lowest points on the surface of the earth; for example, the greatest depth in the ocean is found in the Challenger Deep in the Marianas Trench in the Pacific Ocean, which is approximately 11,000 meters below sea level. In the Atlantic Ocean, the deepest spot is in the Puerto Rico Trench at 8,400 meters.

The trenches are extremely important oceanic features that are closely bound up with the processes that shape the surface of the planet. But there is another feature associated with them that is also tectonically important, and this is the *island arc* system. These arcs are long, curved chains of volcanic islands that are associated with mountain-building processes; they are also the sites of intense volcanic and seismic activity. The arcs are found on the landward side of deep-sea trenches, 200–300 kilometers from a trench axis. The curvature of island arcs is convex toward the open ocean, with a trench on the convex side and usually a shallow sea on the concave side.

It has long been known that earthquakes tend to be concentrated in certain regions, and in the oceans these regions included the mid-ocean ridge system and the trenches. In particular, by the 1920s it was becoming evident that there were a number of earthquake zones (also known as Benioff zones), parallel to the trenches, that were inclined at 40-degree to 60-degree angles from the horizontal and extended several hundred kilometers into the interior of the earth. The plane of the earthquake foci increased in depth under the trenches and toward the continents. Beneath the trenches and their associated features, there was, therefore, a seismically active zone dipping into the mantle at an angle of approximately 45 degrees, to a maximum depth of about 700 kilometers. This is compatible with the sinking of thick, 100-kilometer slabs of cold seafloor under the continents into the asthenosphere (see Figure 6-1). As this happens, the earthquakes are thought to be located in a slab of oceanic crust that is dragged down in a subduction zone because the slab itself remains sufficiently cool to store enough strain to let brittle failure occur in the rocks at such great depths.

There are several lines of evidence that suggest that the sinking of oceanic crust at the trenches is not a simple process. For one thing, the sediment at the bottom of the trenches is often undisturbed, which argues against the direct sinking of material through a trench conduit because such a process should deform the soft sediment. Furthermore, Benioff zone intersects with the ocean floor indicate that downward movement of the oceanic crust occurs landward of the axes of the trenches, rather than directly under them. It was becoming apparent, therefore, that although the theory of seafloor spreading answered many questions, it left others unanswered. Clearly, it was not the geological equivalent of "the theory of everything." However, great steps had been taken along the road to this elusive theory.

▲

There were two major differences between the theories of continental drift and seafloor spreading. First, seafloor spreading did not have the requirement that "light" continents had to plough through "heavy" oceans. Second, seafloor spreading had found a mechanism for the movement of the continents—that is, they didn't need to plough through the ocean floor at all because the ocean floor itself is pulled apart as it spreads, thus moving the landmasses on either side of the opening ocean.

Both mid-ocean ridges and oceanic trenches, so vital to the theory of seafloor spreading, are regions of earthquake and volcanic activity. But it was only *after* the connection between earthquake and volcanic zones, on the one hand, and oceanic ridges and trenches, on the other, had been recognized that the next great leap forward occurred. But before that, the significance of another major feature of the mid-ocean ridge system, the *fracture zone faults,* had to be understood. As this happened, it became clear that the theory of seafloor spreading was only a part of a much greater global process that shaped the entire surface of the earth, not just the ocean basins.

The time had come to attempt to assemble the pieces of the jigsaw into a cohesive, all-embracing process. It was to be called plate tectonics theory.

11

PLATE TECTONICS: THE EARTH SCIENCE "THEORY OF EVERYTHING"

―――――― 🌍 ――――――

A GREAT LEAP FORWARD

The decade of the 1960s was a crucial and extremely exciting time for marine scientists. In 1962, Harry Hess published his seafloor spreading theory, and in 1968 Xavier Le Pichon showed that plate tectonics could describe the evolution of the ocean basins. So what else happened in the pivotal years between these two landmark events to underpin the quantum leap in our understanding of the forces that shape the surface of the planet? Perhaps the best way of untangling the story is to trace its development through a number of seminal papers that were published as the science moved forward.

The ideas that bridged the theories of seafloor spreading and plate tectonics can be divided into two classes. The first concerns movements of the crust associated with the transform faults at the mid-ocean ridges, and the second involves the global distribution of earthquakes and volcanoes.

TRANSFORM FAULTS AND NEW IDEAS

Although it extends for some 60,000 kilometers through the global ocean, the mid-ocean ridge system is a discontinuous feature. The discontinuities are in segments, approximately 10–100 kilometers long, which make the ridge undulate up and down. The segments are offset from each other by *transform faults* that run at right angles to the length of the ridge.

The importance of these transform faults was highlighted by the Canadian geophysicist J. Tuzo Wilson in his 1965 paper "A New Class of Faults and Their Bearing on 'Continental Drift.'" The important point is that one side of the transform fault does not move up or down relative to the other; instead the two sides of the fault slip past each other (see Chapter 14, Figure 14-1c). As this happens, slabs of the mid-ocean ridges are moved along the lines of the faults. Wilson realized that the transform faults were linked to yet a third kind of seafloor movement:

▸ At the spreading ridges, new seafloor is generated and moves across the ocean basins.

▸ At the marginal trenches, old seafloor is destroyed as it moves under the continents.

▸ At the fracture zones on the mid-ocean ridge, slabs of crust slide past each other.

This sliding movement offsets slabs of the seafloor horizontally, out into the whole ocean floor, but unlike the process at the ridges and the trenches, it does so without either creating or destroying crust. Thus, the transform faults provide a link between the spreading ridges and the subducting trenches. This discovery begs the question, *Do transform faults simply play a role in seafloor spreading, or is something else happening here?* This intriguing question was to be answered over the next few years.

Slabs, segments, blocks, plates—all these terms had been used to describe pieces of the earth's crust. Who first came up with the name *plates*? It's difficult to say because the term was bandied about for years outside the formal literature. Perhaps it doesn't even matter, because there is no doubt that an idea was beginning to emerge that just possibly, the surface of the planet was covered with these rigid blocks—whatever they were called.

But one problem that geologists faced was explaining how these rigid blocks, or plates, moved on the surface of a sphere, as indeed they must do in seafloor spreading. This was where Dan McKenzie and his fellow scientist Bob Parker came into the picture.

In the eighteenth century, the mathematician Leonhard Euler had proved that if two rigid bodies move on a sphere, they can only do so by rotating around a single pole. In 1965, E.C. Bullard, J.E. Everett, and A.G. Smith of Cambridge University introduced the concept of an *axis of rotation,* which can be used to describe the movement of blocks of crust on the surface of a sphere. In fact, Sir Edward Bullard's group went a long way to establishing a very necessary component of plate tectonics, i.e., the "fit" of the continents around the Atlantic Ocean, which they demonstrated by converting computer data into a conventional map and using a depth of 2,000 meters rather than shorelines as the edge of the continents. Wegener's drifting continents were no longer just a dream.

In their 1967 paper, *A North Pacific Example of Tectonics on a Sphere,* McKenzie and Parker expanded on the "axis of rotation" concept and demonstrated that if the moving "blocks" of crust were thick enough to actually be rigid, then their motions on a spherical surface would give rise to the kind of seafloor movements recorded by magnetic anomalies. Furthermore, these movements could be explained *"if the seafloor spreads as a rigid plate and interacts with other plates."* They also recognized that if the surface of the earth was, in fact, covered with a series of rigid plates, then those plates must obey strict rules of motion. This meant that they must do one of three things: collide with each other, pull away from each other, or slide past each other. Thus, plate interactions take place only at the margins of the plates and not in the interiors.

In 1968, James Morgan published another of the seminal plate tectonics papers, *Rises, Trenches, Great Faults, and Crustal Blocks.* Like Wilson, Morgan believed that the fracture zones that cut the mid-ocean ridge at right angles to its length, and shift the crest from side to side by hundreds of miles, are a very important third type of seafloor feature—one in addition to ridges and trenches. Following on the "axis of rotation" idea, Morgan proposed that the fracture zones crossing any one mid-ocean ridge are segments of concentric circles that rotate about a common center, or axis, passing through the center of the earth. Morgan saw that two important consequences followed from this.

First, for all the common axes in the oceans and all ridges, trenches, and fracture zones to be involved in one coherent whole, it actually *required* that the crust of the earth be split up into rigid blocks—the plates identified by McKenzie and Parker.

Second, the motions of any two of these blocks, in terms of each other, can be described as rotations around a common axis.

The latter observation was a great breakthrough, because what Morgan suggested was that the best way to accommodate all the different ridge, trench, and fracture zone features in the oceans was to extend the processes that happened on the seabed to *the whole of the earth's crust*—a crust that was broken up into rigid blocks (plates) that moved in the mantle and interacted with each other. In other words, seafloor spreading was only part of a much bigger story. The feature that extended seafloor spreading to a global level, one that involved the whole surface of the planet and not just the oceans, was the transform fault; this was possible because now plates could interact without the constraint of having to create, or destroy, oceanic crust.

According to Morgan, the crustal blocks extend down to the base of the lithosphere where they are underlain by the asthenosphere. Morgan considered how these blocks would interact with each other and confirmed that there are three types of boundaries between them: 1) a rise boundary, where new crust is being created; 2) a trench boundary, where old crust is being destroyed; and 3) a fault-type boundary, where blocks of crust can slide past each other, but crust is neither created nor destroyed. These were later termed *constructive, destructive,* and *conservative* plate boundaries. In Morgan's theory, the earth's crust is divided into a series of blocks (plates) bounded by mid-ocean ridges, deep-sea trenches, large faults, and fold belts. The theory of plate tectonics was becoming in vogue now.

The last big player at this stage of the plate tectonics story was Xavier Le Pichon. He was working at the Lamont Geological Observatory where he combined the theoretical advances made by Wilson, McKenzie and Parker, and Morgan with the large data sets of seafloor magnetic measurements that were then available. In 1968, Le Pichon showed that plate tectonics could describe the evolution of the ocean basins. Furthermore, he used computer analysis to refine the theory and demonstrate that plate tectonics was an integrated system in which, overall, the amount of crust created at the spreading ridges is balanced by the amount destroyed at the trenches.

Also in 1968, a group of American geophysicists—Bryan Issacs, Jack Oliver, and Lynn R. Sykes—showed that plate tectonics could account for a large amount of the earth's seismic activity. What, then, is the connection between seismic activity and plate tectonics? Data had already been gathered that showed that earthquakes occurred in well-defined zones of linear belts (see Chapter 5), and that the areas between the belts are largely seismically inactive. The trick, then, was to explain the existence of these linear belts. The breakthrough came when scientists realized that the belts were aligned the way they were because they defined the boundaries where plates interact with each other and cause tectonic activity to occur. For example, important seismic belts are associated with the seafloor spreading centers at the mid-ocean ridges in the Atlantic and the trench/island arc systems of the Pacific.

Gathering momentum was the idea that the geophysical observations could be explained if the seafloor spreads as a rigid plate *and* interacts with other plates in seismically active regions. It had now been shown that seafloor spreading is not simply a "cause (crustal generation) and effect (crustal destruction)" process, as was at first thought. Rather, the theory had now been racked up from an oceanwide scale to a whole planet scale and had been incorporated into the theory of plate tectonics.

PLATES AND "PLATE TECTONICS"

The theory of plate tectonics relies on the fact that the planet earth is continuously in motion; both on the surface where plates swim around, and in the interior where convection cells operate.

According to plate tectonics theory, the surface of the earth is covered by a series of interlocking lithospheric plates on which the continents and the ocean basins are carried. The plates are relatively large, rigid "rafts" that are decoupled from the underlying mantle by the softer, plastic, asthenosphere on which they float like icebergs. They are in continual push-and-pull motions, usually as single units, in response to forces acting on their boundaries, which are the sites of seismic activity.

It is generally agreed that there are seven or eight major plates:

the Eurasian, the North American, the South American, the Indo-Australian (sometimes separated into Australian and Indian plates), the Pacific, the African, and the Antarctic plates. There are also a number of intermediate plates that include the Nazca, Cocas, Arabian, Scotia, and Philippine plates, together with several minor plates. The thickness of the tectonic plates varies, but away from the margins they are usually between 70 and 200 kilometers thick.

The distribution of the plates is illustrated in Figure 11-1, and the plate boundaries should be compared to the global distribution of earthquake zones (see Chapter 15, Figure 15-3). It is important to realize that although the plate boundaries are sites of seismic *and* volcanic activity, they do not usually coincide with the boundaries of either the oceans or the continents. In fact, with the exception of the Pacific Plate, the major plates have both oceanic and continental components. Furthermore, the behavior of plates is not strongly influenced by whether they carry oceans or continents.

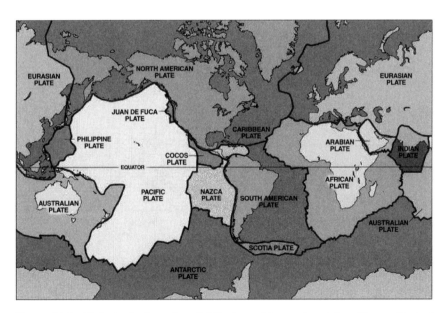

Figure 11-1. Plate tectonics—the distribution of the major plates. The surface of the earth is covered by a series of plates with a thickness of between 70–200 kilometers. The plate boundaries do not usually coincide with the boundaries of either the oceans or the continents—in fact, with the exception of the Pacific Plate, the major plates have both oceanic and continental components. Credit: S. Nelson, "Global Tectonics."

Convergent Plate Boundaries

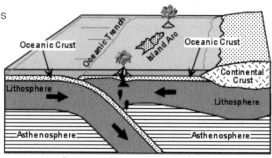

Ocean—Ocean Convergence

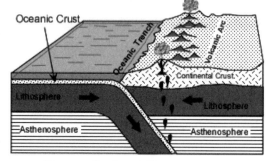

Ocean—Continent Convergence

Divergent Plate Boundaries

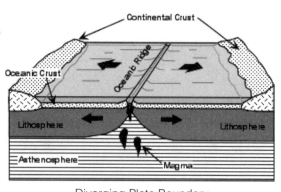

Diverging Plate Boundary
Oceanic Ridge—Spreading Center

Figure 11-2. Interactions at plate boundaries. The earth's surface is covered by a series of interlocking lithospheric plates on which the oceans and continents are carried. The plates are essentially rigid, and most active deformation takes place at their boundaries during interaction with other plates. The plates move relative to each other and come into contact under a number of different boundary conditions that govern whether they move apart (divergent boundaries), come together (convergent boundaries) and collide, or slide past each other (transform boundaries). Credit: S. Nelson, "Global Tectonics."

However, seafloor spreading is embedded in plate tectonics theory, and particular attention should be paid to the condition when a plate boundary exists *between* a continent and an ocean, because here subduction can occur, resulting in intense seismic activity.

Most active deformation of tectonic plates takes place at their *boundaries* during interaction with other plates. The plates move relative to each other and come into contact under a number of different boundary conditions that govern whether the plates move apart, come together and collide, or slide past each other (see Figure 11-2). The manner in which the plates collide also dictates the kind of activity taking place—activity that can result in mountain building, rifting, and earthquake and volcanic activity, and so is able to explain many of the great features on the surface of the earth. What happens at plate boundaries is therefore crucial in shaping the surface of the planet.

Divergent, or Constructive, Plate Boundaries. They are found mainly at the mid-ocean ridges where the plates are forced apart as new seafloor is generated. In this way, the divergent boundaries are crustal sources. The mid-ocean ridge system is topographically high because the region below, where the new crust emerges, undergoes thermal expansion as pressure in the mantle is released. This process results in an elevation of the seabed that helps the new crust slip apart—sometimes called a "ridge push."

It is important to realize that divergent plate boundaries are not confined to the middle of the oceans, but can also be found where a continental plate breaks up. In this way, the rifting that caused the Great Rift Valley in Africa originated with the initial separation of the Arabian from the African tectonic plate, and is currently splitting East Africa away from the rest of the continent along a spreading axis. The breakup of a continent in this manner is thought to be associated with upward movement caused by convection in the mantle. As a result, the crust thins and becomes domed above the rising plume, processes that are accompanied by volcanic activity and faulting. Close to the spreading axis, the faulting produces rift valleys as blocks sink down to form rift zones, or *grabens.* As the process develops, the rift valley extends in length and may eventually find an outlet to the ocean as a new narrow linear sea, as happened in the formation of the Red Sea. Not all rift valleys, however, develop into spreading centers.

Convergent, or Destructive, Plate Boundaries. These are subduction zones where crust is being destroyed. In these regions of trenches and island arcs, one plate moves beneath another at a 45-degree angle and sinks into the mantle. Thus, convergent boundaries are crustal sinks, and when the ocean floor undergoes subduction into the underlying mantle, it takes the sediment cover with it. Convergent plate boundaries are formed when two plates collide, and there are a number of possible types of collisions:

1. Collisions between oceanic and continental plates. Here, the denser oceanic crust is forced by *subduction* into the mantle below the lighter continental crust, via the conduit of a deep-sea trench. In fact, the continental crust is too light (i.e., has too much buoyancy) to sink into the mantle at all. That's why the continents are much older than the oceans, which are continually being recycled back into the mantle.

During subduction, a slab of lithosphere, consisting of oceanic crust and associated sediment cover, descends into the mantle forming a deep-sea trench. As this happens, shallow earthquakes are generated along the upper surface, and at depths greater than 95 kilometers, heat drives off the water and other volatiles from the subducted sediments into the mantle, where they act as a kind of "flux" and induce partial melting of the mantle material. The partial melt is less dense than the rest of the mantle and will rise, either to gather as "ponds" beneath the continental crust, or to break out and result in volcanic activity. The volcanoes can be created in a line that runs parallel to a trench, which in an oceanic setting gives rise to volcanic islands. Some of the sediment on the descending plates can be folded and faulted and thrust against the continental slab to form mountain chains, such as the Andes, which result from the Nazca Plate (oceanic) colliding with, and being subducted under, the South American Plate (continental). This picture of subduction is considerably simplified, and the actual dynamics of the process, which can involve features such as volcanic arcs, forearc and backarc basins, and the "scrapping off" of accretion sediment wedges, are more complex and are by no means totally understood. From our purposes, however, the simplified explanation will do.

2. Collisions between two oceanic plates. They result in one plate being thrust beneath the other, which rides over it. This kind of collision gives rise to volcanic activity and may lead to the formation of an island arc—such as the Aleutian Islands. When two oceanic plates collide and subduction occurs, the uppermost plate can become stretched and lead to the formation of basins—an example is the Lau Basin in the South Pacific, which is forming above the Tonga subduction zone.

3. Collisions between two continental plates. They give rise to a *collision zone,* in which the major motions are restricted to one plate overthrusting, or crumbling under, the other. For the collision between two continental plates to occur, the plates must have been separated by an ocean, and as they move toward each other they displace the ocean by subducting the seafloor under one of the continental plates. This results in the formation of volcanic arcs on either of the two continental plates, or in the development of an island arc in the ocean section—a feature that owes its curved shape to the fact that the subducting slab is penetrating into a sphere. The remaining seafloor is then consumed and the continents collide. These collisions, where continents fuse together, can result in the formation of great high mountain ranges such as the Himalayas, which were generated as the Indian Plate rammed into the Eurasian Plate. The Alps and the Urals are other examples of mountain ranges resulting from continent/continent collisions.

Conservative, or Transform, Plate Boundaries. Transform faults link the divergent and convergent margins as part of the plate tectonic network that divides the crust into plates. At conservative plate margins, material is neither created nor destroyed; instead, the plates move laterally and slide past one another along transform fault boundaries, or fracture zones—most of which are found at the mid-ocean ridges at intervals of 95 kilometers. However, just because transform faults slide past each other does not mean that they are not involved in seismic activity as they move around. In fact, they cannot simply slide past each other without any reaction because of the friction generated, which leads to the setting up of strain. The release of this strain leads to the generation of earthquakes, which can define plate boundaries. This is amply demonstrated by the earthquake activity around the San Andreas Fault in California—an example of an on-land transform fault.

Finally, it should be stressed that individual plates can have combinations of these various boundaries—divergent, convergent, and transform—at different places on their surfaces.

THE THEORY OF EVERYTHING?

Perhaps.

Plate tectonics was certainly the first theory to come close to offering a coherent explanation for the formation of all the features on the surface of the earth—features such as oceans, continents, trenches, and mountain ranges. Plate tectonics can also rationalize earthquake and volcanic activity—as we shall see later. And everything happens at the plate margins, the great engines that drive geological processes.

The environmental impact arising from the motions generated by plate tectonics can be divided into two categories.

The first impact involves long-term effects that shape the topography of the entire planet—for example, the movements of the continents and the birth and death of the ocean basins. These events can be described in terms of what has become known as the "Wilson Cycle" (named after Tuzo Wilson). Although it is now considered to be an oversimplified concept that cannot account for some detailed geological features, the Wilson Cycle is a useful way of understanding the broad brushstroke history of the planet. Essentially, the history of the earth is interpreted in terms of the cyclic opening and closing of the ocean basins through an evolutionary sequence that involves a series of theoretical stages. In the embryonic stage, continental rift results in the formation of a rift valley (e.g., the African Great Rift Valley), which is followed by the generation of a young narrow ocean (e.g., the Red Sea). The ocean then expands as the continents move further apart, and new ocean floor is generated at active spreading ridges (e.g., the Atlantic). Since continued expansion is not possible, the next stage involves the development of destructive margins where subduction occurs via deep-sea trenches. When the rate of subduction exceeds the rate of the production of new crust, shrinking of the ocean basin takes place (e.g., the Pacific) as the plates adjust to preserve planetary equilibrium. In the final throes, as oceanic shrinking continues, collisions occur. Collisions between two continental plates, originally separated by an ocean, can result in uplift and the building of great mountain ranges (e.g., the Himalayas). Eventually, the ocean closes, and the cycle is then repeated when continental rifting is restarted.

The second impact encompasses short-term effects that can disturb the environment to such an extent that their impact on life-forms is immediate and, indeed, sometimes catastrophic. Included in this category are earthquakes and volcanic activity, and the secondary generation of tsunamis. We will look at these short-term environmental impacts in detail in later chapters. Before that, it is fitting to conclude the discussion of the great theories that have attempted to explain how the planet evolves by homing in on a feature that is a pivotal consequence of plate tectonics—the marine hydrothermal system.

At this stage, however, we can answer one of the questions we posed as we began the journey toward plate tectonics theory: *How is it possible that the continents, the solid earth we live on, can drift around the surface of the planet?* The answer is that the continents drift around the surface of the earth because they ride on large tectonic plates that are continually in motion.

The jigsaw is complete.

Except for one important question: We now understand that the configuration of the continents has changed with time (continental drift), but what started the drift in the first place? Perhaps, as James Morgan suggested after he had written his seminal paper, the answer lies in "hot spots" that appear to have been associated with the generation of rift valleys. Hot spots are mantle plumes, with a large head and thin tail, forming volcanic centers below the lithosphere that appear to remain in a fixed position and are supplied with magma from deep in the mantle. However, some geoscientists now think that hot spots may be neither located deep in the mantle nor in a fixed position.

The best known hot spot is probably the one associated with Hawaii (see Chapter 13), but others are found beneath Iceland, the Azores, and the Galapagos Islands and, in a continental sitting, under Yellowstone Park in Wyoming; altogether around 50 presently active hot spots have been identified. Hot spots have received a good deal of attention over the past decade, and as well as supporting volcanic activity in island locations, for example, Hawaii, they are now thought to interact with the mid-ocean ridge system by acting as magma suppliers, such as in Iceland.

According to the American marine geologist Roger Anderson, hot spots could act as blowtorches to "cut the continents apart."

12
HYDROTHERMAL ACTIVITY: A NEW OCEAN WORLD

THE THIN RED LAYER

It had been known for many years that red-brown colored sediments rich in iron and manganese can be found in association with the mid-ocean ridge system, and as far back as the *Challenger* expedition, metal-rich sediments had been found in the area of the East Pacific Rise. But the origin of the iron and manganese in the sediments wasn't known until the mid-1970s, when metal-rich brine solutions were found in an area of active seafloor spreading in the Red Sea.

A new ocean is being born in the Red Sea. The African Rift Zone is a down-dropped fault block constructed along a divergent boundary and the Red Sea, which lies in a narrow linear strip between the African and Arabian tectonic plates and occupies part of this zone of depression and faulting. Here, the rifting, which started about 30 million years ago, has progressed to the stage at which a continental split has developed and a new ocean is being opened at a rate of about 1.5 centimeters a year.

Hot brines had first been found in the Red Sea by Russian scientists in the 1880s, and over fifty years later by Swedish oceanographers on board the research ship *Albatross*. At first, it was thought that these brines had been formed from the evaporation of seawater that had become dense enough to sink to the bottom of the water column. This thinking changed, however, in the mid-1960s, when the British research ship *Discovery*, and a year later the U.S. research ship R/V *Atlantis 2*, collected and analyzed samples of these brines. The high-salinity (up to 250 parts per thousand, or ppt) compared to that of 30–35 ppt for most seawaters), high-temperature (40°C–60°C) brines were found at a depth of about 2,000 meters in a series of depressions in the central rift area of the Red Sea in 1965. The brines had one other extremely important characteristic: Relative to normal seawater, they were considerably enriched in manganese and iron, and the so-called heavy metals (e.g., lead, copper, zinc). It was suggested that the metals had been leached from (i.e., dissolved out of) the underlying rocks by the high-temperature geothermal waters. A sample of the seabed surface sediment collected in the vicinity of the brines had the appearance of a hot, tar-like ooze. When more sediment samples were taken later, they showed the presence of different colored layers, each rich in specific metals like iron, manganese, zinc, and copper. It was then proposed that after they had been leached from the underlying rocks, the metal-rich sediments were precipitated out of the geothermal brines on contact with seawater.

Data obtained from the Red Sea brines provided an extremely important step forward in our understanding of the formation of metal-rich sediments. For instance, the findings threw light on some of the processes that take place when hot, metal-rich solutions interact with seawater. David Cronan, a British geochemist working out of Imperial College, London, used the brines as a model to outline the sequence of events in which components precipitated out of the hydrothermal solutions. This sequence had the following order: 1) *sulfides*, for example the minerals sphalerite (zinc, iron sulfide), pyrite (iron sulfide), marcasite (iron sulfide), galena (lead sulfide); 2) *iron silicates*, for example the minerals smectite and chamosite; 3) *iron oxides*; and 4) *manganese oxides*.

Then, in 1969, the Swedish geochemist Kurt Bostrom and his coworkers produced a map of the global ocean showing that metal-rich deposits were present on all the major mid-ocean ridges where new crust was being generated. Their work provided a strong indication that metal-rich sediments may be tied in to processes occurring during seafloor spreading, although at the time, the reasons were not fully understood.

THE LOST HEAT

Heat escaping from the seafloor by *conduction* should be highest in places where hot magma erupts under the seabed to produce new crust. But, as was pointed out in Chapter 8, in fact, the heat flow out of the seafloor along parts of the mid-ocean ridge system was considerably lower than expected. This fact led geoscientists to the conclusion that perhaps some as yet unknown mechanism, in addition to conduction, must be cooling the newly formed oceanic crust by carrying away heat.

Gradually, the idea began to emerge that perhaps the mechanism that was carrying this heat away was the production of hot springs formed during hydrothermal venting activity at the centers of seafloor spreading—a process that has similarities to boiling geysers on land. But the question of whether hydrothermal activity actually does occur on the oceanic spreading ridges would only be answered definitively when the ridge environment could be studied at close hand.

EXPLORING THE MID-OCEAN RIDGES

Although a variety of high-tech instrumentation had allowed oceanographers to photograph the seabed, map it, take samples of it, and probe it for various kinds of seismic and heat flow data, no one had actually been able to go down and observe the deep-sea firsthand.

The key step in achieving this goal was the development of the manned submersible. The most famous of these vessels was the *Alvin*, a self-propelling, deep-sea pod owned by the U.S. Navy and operated out of the Woods Hole Oceanographic Institution. *Alvin*, which was commissioned in 1964, could carry a crew of three to depths of 4,000 meters. The submersible had an array of sophisticated instrumentation, including lights, video cameras, computers, and highly maneuverable sample-collecting arms to bring in geological and biological samples. Chemical measurements could be made using a wand that was used to operate thermometers, electrodes for taking real-time chemical measurements, and a water sampling system.

There were a number of landmarks in the vessel's history. The first came in 1974 with *Alvin*'s participation, together with the French submersibles *Cyana* and *Archimede,* in Project FAMOUS, a French-American expedition to investigate an area at 37°N on the Mid-Atlantic Ridge, some 320 kilometers southwest of the Azores. FAMOUS employed state-of-the-art instrumentation; for example, detailed maps were compiled from the use of narrow-beam and multibeam echo sounding systems, and it was the first time that scientists had viewed the mid-ocean ridge firsthand and up close as they descended over 1,500 meters into the abyss. On the basis of finding that the rift valley in the ridge crest is created by large faults in the newly formed oceanic crust, FAMOUS went a long way toward proving seafloor spreading was a viable theory.

But hydrothermal vents were not found during the FAMOUS expedition. That discovery had to wait for another *Alvin* dive in 1977; this time to the Galapagos Ridge, on a day that changed forever our view of the oceans.

A NEW WORLD

Imagine the scene as geologist Jack Corliss and geochemist John Edmonds looked through the *Alvin* portholes onto the Galapagos Rift, an undersea volcanic ridge 320 kilometers west of Ecuador in the Pacific Ocean. The submersible had been dropping through a world of total darkness for more than two hours as it descended more than two kilometers of water. Then, as the seabed approached, the searchlights were switched on to reveal one of the most important discoveries in the entire history of earth science: a strange new world that existed at the bottom of the sea.

As the scene unfolded, the scientists saw an environment shimmering in heat. It was a world of strange life-forms living around a bubbling spring venting onto the seabed. There were reefs of muscles, fields of giant white clams, and perhaps the most astonishing of all, thickets of giant tube worms, up to eight feet tall. The tails of the worms were planted on the seafloor, and blood-red plumes at their heads swayed in currents like fields of poppies.

It was a world of striking color at the bottom of the drab colorless sea, an oasis in a barren desert with all the colors of life. Except one. Green. For this was a world that flourished in the absence of sunlight. A world without photosynthesis.

THE PIECES BEGIN TO FIT TOGETHER

The presence of hydrothermal solutions had been identified in the Red Sea, but it was only when *Alvin* dived to the seabed at the Galapagos Rift that they were found at a major mid-ocean spread-

Earthquake fault. Most earthquakes are associated with faults. Here, a fence has been ruptured during earthquake faulting causing horizontal displacement. *The U.S. Geological Survey.*

Pyroclastic flow. Dense clouds consisting of gases and small particles of lava, pyroclastic flows, also called nueé ardente, are one of the most dangerous "along the ground" volcanic threats, and can leave devastation in their wake. In this eruption of Mount St. Helen, a column of ash and pumice (darker colored) is ejected vertically into the air from the crater, whereas a pyroclastic flow (lighter colored) hugs the ground and begins to move down the mountain at speed. *The U.S. Geological Survey.*

The birth of an island. Volcanic activity at the North Atlantic mid-ocean ridge spreading center gives rise to the island of Surtsey, Iceland. The image of the ash cloud was taken in 1963, three days after the first sighting of the island. *The University of Colorado; image obtained from NOAA.*

Hydrothermal vent. "Black smoker" at the mid-ocean ridge spreading center in the Atlantic Ocean. The chimney is debouching a hydrothermal fluid rich in clouds of precipitated black sulfides. *P. Rona/NOAA.*

Earthquake damage. The town of Balakot was severely damaged in the 2005 Pakistan earthquake. This earthquake, which occurred in October, illustrated the difficulties in providing aid to remote, mountainous regions as winter approached. *Gregory Takats, AusAID.*

Earthquake damage. In the town of Muzaffarabad, all schools and most medical centers were destroyed in the 2005 Pakistan earthquake. *Gregory Takats, AusAID*

Tsunami damage. Devastation caused by the 2004 Indian Ocean tsunami, waterfront area, Banda Aceh, Indonesia. *J. Borrero, University of Southern California, Viterbi School of Engineering Tsunami Research Center.*

Tsunami damage. Boat lifted onto a house during the 2004 Indian Ocean tsunami, Banda Aceh, Indonesia. *J. Borrero, University of Southern California, Viterbi School of Engineering Tsunami Research Center.*

ing center. Even then, the temperatures of the fluids were not especially hot, and it wasn't until later that hydrothermal vent fluids with temperatures greater than 350°C were found at various sites on the East Pacific Rise. Only when these high-temperature fluids had been discovered could the story behind the mid-ocean ridge hydrothermal venting system be fully unraveled.

The hot fluids begin life at the spreading centers as cold seawater that percolates down into the newly formed crust via cracks and fissures. It then begins to circulate below the surface and comes into contact with hot, newly formed rock. This initiates reactions between the water and rock that give rise to hot, acidic, reducing solutions. During the reactions, some components are leached out of the rock, and some are taken up by the rock, with the overall result that the composition of the original seawater is considerably modified. This reaction affects both major and minor elements, gases and the isotopic ratios of some elements, with the result that the hydrothermal fluids have a very different composition from that of seawater.

When it was found that the hydrothermal venting system was a global feature of the mid-ocean ridge, scientists realized that the hot springs provided a mechanism for the transfer of heat from the solid earth to the liquid ocean. In the process, cold seawater penetrates downward and is heated rapidly. The buoyant hot fluid is then forced back to the surface of the seabed, where it emerges through vents as hydrothermal springs, carrying heat with it.

The "lost" heat had been found.

DIVERSITY: "BLACK SMOKERS" AND "WHITE SMOKERS"

The strongly acidic hydrothermal solutions, which are rich in hydrogen sulfide and heavy metals and poor in oxygen, are vented through the ocean floor where they come into contact and mix with cold (2°C), ambient, normal seawater. This mixing of the two water end-members switches on the precipitation sequence, as outlined previously for the Red Sea, which operates in the order:

Sulfides ➡ iron silicates ➡ iron oxides ➡ manganese oxides

The critical factor with respect to the kind of vent "plumbing" (i.e., the type and composition of the chimneys out of which the fluids emerge onto the seabed) and the temperature of the venting fluids themselves is the location at which the mixing of the two end-member waters—venting fluid and seawater—takes place. To date, two principal hydrothermal venting systems have been found on the seabed. They are the so-called *white smokers* and *black smokers*, named after the type of vent fluid they discharge into seawater:

▸ *Low-Temperature Hydrothermal Systems.* They were first found on the Galapagos Ridge during the 1977 *Alvin* dive. Here, the fluids, which had temperatures in the range of 6°C to 10°C, were venting milky-colored plumes from the seafloor at a rate of 2–10 liters per second. It was suggested that the hydrothermal solutions had mixed with seawater at depth within the venting system and so were fairly cool when they reached the sea surface. This, in turn, meant that the precipitation sequence had been initiated at depth. As a result, in the "white smoker" hydrothermal systems, the sulfides and sulfide-associated metals are taken out of solution below the surface, and the vent fluids are whitish in color.

▸ *High-Temperature Hydrothermal Systems.* They were first found on the East Pacific Rise, part of the mid-ocean ridge in the Pacific. In these systems, the venting fluids exit through mounds of sulfide-rich debris; in the most dramatic form, the fluid jets from tall (up to 15 meters high) black chimneys on the seafloor. In the high-temperature systems, hot, acidic, reducing sulfide and metal-rich solutions, with temperatures of up to 350°C, are debouched as black plumes directly into the bottom waters through the chimneys, which are composed of the sulfides of metals such as iron, zinc, and copper. These are the "black smokers," so-called because of the black color caused by the precipitation of fine-grained sulfides as the venting fluids are mixed with seawater, giving the appearance of venting smoke. Black smokers can grow rapidly, up to 20–30 centimeters a day.

The initial mineral that goes to build a black smoker chimney is probably anhydrite (calcium sulfate), which is precipitated during the mixing of seawater and hydrothermal fluids. The anhydrite forms a ring structure around a jet of hot (350°C) vent fluid, and then begins to build a thin chimney wall. This process is followed by the sequence in which sulfides, silicates, and oxides are precipitated, some in the interstices of the chimney walls, some in halos away from the chimneys. In black smokers, therefore, the precipitation has only been initiated at the seabed and the full sequence of precipitates is present. The hydrothermal sulfides are generally confined to the chimneys or the immediate vicinity of the venting system. Red-colored iron and manganese oxides, however, which are characteristic deposits of hydrothermal venting at the spreading centers, can be found over a much greater area as halos around the chimneys.

▲

The secret of the red-brown layer had been revealed—it was a hydrothermal precipitate.

It is now known that a metal-rich red sediment consisting chiefly of iron and manganese oxides is formed at most of the spreading centers. It follows, then, that since new crust is moved away from the ridges during seafloor spreading and is blanketed with sediment as it does so, the red-colored metal-rich deposit should lie at the base of the sediment column in *all* locations. In other words, it should be the first sediment ever deposited on the seafloor. This hypothesis was confirmed by data obtained from the DSDP. When the full sediment column was sampled, it showed that a thin layer of the red-brown, metal-rich deposit was present at many oceanic locations. The thin red layer on the basalt basement could now be incorporated into models of marine sedimentation (see Chapter 9, Figure 9-1).

Hydrothermal venting systems have been found at the spreading centers in all the major oceans, and they are often concentrated in specific fields. For example, in a 15-kilometer portion of the Endeavour Segment of Juan de Fuca Ridge in the Pacific Ocean, five active hydrothermal vent fields have been reported, one of which contains more than 100 active black smokers. Vent fields are often given exotic names, like Hole to Hell, Tube-Worm Barbecue, Valley of the Shadow of Death, Rose Garden, and Snake Pit.

The discovery of the "white smoker" and "black smoker" types of plumbing demonstrated that not all hydrothermal venting systems were the same. This view was strengthened during subsequent years when it became apparent that there was considerable diversity among individual vent fields brought about by factors such as venting fluid temperature, the depth to which the fluids had penetrated below the seabed, the type of rock encountered, and the distance traveled before a fluid exited on the seafloor. In fact, some systems changed character on very short-term geological timescales, ranging from days to years.

The diversity of hydrothermal systems was highlighted by the discovery of the Lost City vent site at 30°N in the Atlantic. This pale, ghostly environment was discovered on a seamount (i.e., undersea mountain) some considerable distance (15 kilometers) from the ridge axis on old, rather than new, seafloor that is apparently unsupported by active volcanic activity. At this site, the fluids were vented through tall lime-gray carbonate chimneys—at about 60 meters, the highest then found on the seabed. There are a number of important differences between the vent system found here and the system found on the East Pacific Rise. For example, the Lost City chimneys are made of carbonates, not sulfides, and they vent white alkaline fluids at temperatures that reach around 90°C, instead of black acidic fluids at 350°C.

Some marine biologists believe that the Lost City vent field may be the closest marine venting system to the hot springs found during the early history of the earth, and it may hold clues into the dynamics of the evolution of life on the planet.

VENT COMMUNITIES

Could there be a harsher place on the planet for the sustenance of life than the hydrothermal springs on the deep-sea bed? On the face of it, everything seemed stacked against finding any living creature at a deep-sea hydrothermal spring. Yet, in this world of extreme high temperature, high pressure, and pitch darkness, in the presence of hot, acidic, metal-toxic solutions full of poisonous gases, life was indeed found. Against all the odds, a rich, varied community of extraordinary life-forms exists in the shadow of the hot springs.

Since *Alvin's* first dive to the site of a vent system, more than 300 species have been discovered around the global mid-ocean ridge hydrothermal system, and incredibly, 90 percent were new to biology. The life-forms most commonly found around the hot springs include fish, giant clams, giant white crabs, muscles, limpets, blind shrimp, anemones, octopuses, and various worms. Smaller creatures, such as miniature lobsters, sand fleas, and amphipods, together with bacteria, also inhabit the hot spring environment. But perhaps the most striking, and certainly the most biologically interesting, members of the vent community are the giant tube worms known as *vestimentiferans*. Up to a meter and a half long and 30–40 millimeters wide, they have blood-red plumes at their heads and live in tubes up to 3 meters in length. For as long as the vents exist, usually up to a few years, the specialized animal communities continue to inhabit this harsh environment.

How far we have come from the pre-*Challenger* days when, in 1843, naturalist Edward Forbes told a meeting in London that it was probable that nothing was alive on the whole deep-sea floor. In fact, Forbes went even further; on the basis of the rate at which life-forms decreased with depth, he postulated that below 300 fathoms (about 550 meters), the ocean was a vast lifeless *azoic zone*. Well beyond Forbes's time, scientists continued to believe that the ocean bed was a dark cold desert, with the only supply of food being the debris of plankton and, on occasion, the carcass of a larger animal that sank down the water column.

The food chain in the ocean is dependent on photosynthesis. Phytoplankton, minute plant organisms that live in the upper water column, grow by photosynthesis and can be thought of as the grass of the sea. They are fed upon by zooplankton, which by the same analogy may be considered the cattle, or grazers, of the sea. In turn, the zooplankton are consumed by larger predators, and so the food chain progresses. This entire chain is underpinned by photosynthesis. Indeed, almost all life on earth depends, directly or indirectly, on the solar energy required for the process of photosynthesis. This is the synthesis by green plants (phytoplankton in the oceans) of carbohydrates from water and carbon dioxide with the aid of sunlight. However, the communities around the hydrothermal hot springs are far removed from the influence of solar energy, and any food chain has to depend on another source.

Vent communities presented a biological enigma—until it was found that the answer lay in bacteria. Bacteria are the first organisms to sense new hydrothermal vents and colonize them in "bacterial clouds." Bacteria are unicellular microorganisms that can use chemical energy for their metabolism in the process of *chemosynthesis*. In a sense, this is similar to photosynthesis, since both processes synthesize carbohydrates from carbon dioxide. However, whereas in photosynthesis the energy required is supplied via sunlight, in chemosynthesis the source of energy is a chemical compound; in the case of the vent communities, hydrogen sulfide.

The hydrogen sulfide is a reduced sulfur compound, and when it is oxidized the energy released is used to synthesize organic matter, the molecular oxygen needed to drive the process being provided by seawater. In some instances, however, ammonia can be used as the energy source.

Once established, the bacteria grow into a thick mat that attracts other organisms. This is the initial step at the base of the vent community food chain in which some life-forms, such as mollusks, feed directly on the bacteria. Other predators fed on the animals that have utilized the bacteria. Still others, such as the tube worms, allow the bacteria to live in their bodies and supply nutrients directly to their tissues. In the deep-sea vent field, therefore, life depends on chemicals from the earth rather than energy from the sun.

The red-plumed tube worm is a particularly interesting vent community life-form because it lacks its own digestive system. It survives by endosymbiosis, i.e., essentially one organism living inside the body of another in a way that has benefits for both parties. The insides of the worms are packed with millions of bacteria, and the red plumes at the top of their bodies are filled with blood. The blood contains hemoglobin that binds hydrogen sulfide from the vent fluids and transports it to the bacteria living inside the worm. The bacteria then oxidize the hydrogen sulfide and convert carbon dioxide into carbon compounds that act as food for the worm.

There is a complication to the hydrothermal system, however. A dim glow of unknown origin has been found at some hydrothermal vents. This provides a source of natural light on the dark seafloor and raises the question, Could photosynthesis, as well as chemosynthesis, be possible at the vents?

SAFE HAVENS AND THE ORIGIN OF LIFE

Perhaps the greatest controversy arising from the discovery of the mid-ocean ridge hydrothermal system concerns the role it plays in the origin of life. This was a topic that fascinated Jack Corliss from the moment he first looked upon the hydrothermal vents on the

Galapagos Rift that day in 1977. For years many scientists believed that life on earth started when cellular organisms appeared in a "biochemical soup" in a tidal pool, or freshwater pond, in the presence of sunshine. But the finding of the oceanic hydrothermal vent system was to challenge this view.

It is generally agreed among evolutionary biologists that two important criteria have to be met to produce life. One is that the life-form must be able to replicate itself, and the other is that it should carry an information code that can be passed to future generations in order to give life continuity. DNA (deoxyribonucleic acid) is the storage unit for the genetic information, and RNA (ribonucleic acid) is the dispenser unit for it. Originally, it had been thought that replication involved DNA, RNA, and enzymes in protein, although it is now believed that initially a kind of RNA, functioning as an enzyme, could replicate itself and only later did DNA and protein evolve to help in the process. But the RNA, which is simpler than bacteria, must itself have originated somehow. So what were the precursor chemicals—the first organic compounds?

The chemicals from which life emerged are thought to have been methane, ammonia, hydrogen sulfide, carbon dioxide or carbon monoxide, water, and phosphate. Serious work on how these chemicals combined to form the building blocks of life only really started in the 1920s and 1930s, when the Russian biologist Aleksandr Oparin, who is often referred to as the "Darwin of the twentieth century," suggested in his book, *The Origin of Life on Earth*, that the reactions were initiated in the primitive reducing atmosphere of the earth. However, it was not until 1953 that the experiment proving that organic chemicals could be synthesized from inorganic starting material was carried out. This was the famous Miller-Urey experiment in which water, methane, ammonia, and hydrogen were sealed inside a connected system of glass vessels and subjected to a simulated atmospheric lightning attack using electrodes. The result was that between 10 percent to 15 percent of the carbon was turned into simple organic compounds that included amino acids. Thus, although the experiment did not produce life itself, it showed that the biochemical building blocks that were an essential step in the process of starting life could form naturally in the experimental mixture. Put another way, if the building blocks fell into water, a *prebiotic soup* would be created that contained the building blocks of life and could lead to the formation of the first living cell.

The conditions used in the Miller-Urey experiment were criticized by a number of scientists for not truly representing the atmosphere of the early earth. In particular, the atmosphere back then was thought to be more oxidized than in the experiments, which is important because it is harder to form organic compounds in an oxidized atmosphere than in a reduced hydrogen-rich one. It was believed that life had originated under reducing conditions, so when it was suggested that the atmosphere of the primitive earth was not reducing, attention turned to other suitable environments—which included outer space and the deep-sea hydrothermal venting system.

The idea that the deep-sea hydrothermal venting system could have been the birthplace of life had one attractive advantage over the Miller-Urey hypothesis. The conditions on the surface of the early earth were, to say the least, extremely hazardous as the planet was bombarded continuously with cosmic debris. True, deep-sea hydrothermal vents had their own dangers, but they had one important requirement for the preservation of life if it did indeed originate in those furnace conditions early in the history of the earth: The deep-sea vents were protected by the overlying water column and could provide a continuity of environment. In fact, some biologists believe that the bottom of the deep-sea could have been the *only* environment on the planet that could have offered a safe haven. This provision of a safe haven where a life-forming process can be protected, to operate in a continuous undisturbed fashion for a long period of time, is central to the theory proposed by Jack Corliss at the Twenty-Sixth International Geological Congress in Paris in 1980.

Essentially, Corliss postulated that the process involving life-forming reactions began at depth in the hottest part of the vent system where seawater had filtered down to encounter hot rock that it cracked open. In the high temperatures of this furnace (up to 1,300°C), elements such as carbon, oxygen, hydrogen, nitrogen, and sulfur interacted to form organic compounds. But the question is, how could the compounds have possibly survived at such great temperatures? The answer, according to Corliss, lies in the strong temperature gradient found in vent chimneys, where the temperature of the superheated water (greater than 300°C) is lowered by newly circulating seawater that has a temperature of approximately 2°C. As this happens, the organic compounds are quenched almost as soon as they are formed, thus preserving them.

In the next stage, the organic compounds flow around the vent system, rising up the fractures in the cooled rock to emerge in the chimneys at the seabed. Instead of being vented, however, and dispersed in a vast oceanic world in which they may never encounter another organic compound, they attach themselves to clay minerals in the linings of the chimneys, where colonies of them develop. Here, it was suggested that the organic compounds interacted with each other over long periods of time until life-forming compounds appeared in these "safe havens."

The hydrothermal vent "origin of life" theory also provides a means to the next stage in the evolution of life—namely, self-replication. To achieve this, it has been proposed that clay mineral crystals in the vent chimneys do not simply act as a surface to which organic compounds can attach themselves and so prevent them being spewed out. Rather, the clays are thought to play a much more fundamental role in the process that changes the simple organic molecules into self-replication entities; they act as catalysts for the reactions involved. Eventually, the thinking goes, this led to the formation of self-replicating RNA, which then left the clay mineral home.

Another insight into how life may have originated in the oceanic hydrothermal system came with the innovative work of Gunter Wachtershauser, a German scientist who produced a detailed step-by-step blueprint to describe how the earth's oldest raw materials might have given birth to life. The most significant aspect of Wachtershauser's work was that the evolution of biochemical pathways was fundamental to the start of life. He devised an assembly-line approach operating at the ocean floor to transform raw inorganic chemicals, such as carbon and hydrogen, into the biological molecules that become the building blocks of life. Wachtershauser's work has been criticized on the ground that his experiments did not incorporate the scorching temperatures or the high pressures that are features of the vent system—although other workers are now testing the hypothesis at more extreme conditions.

There have been other attempts to support marine hydrothermal vents as life-originating environments. In 2002, for example, Professor William Martin of the University of Düsseldorf and Dr. Michael Russell of the Scottish Universities Environmental Research Centre in Glasgow came up with the idea that life originated in small compartments in iron sulfide rocks in vent systems that acted as organic incubators. These sulfide cavities provided the right microenvironment in which the building blocks of life— hydrogen sulfide, hydrogen, and carbon monoxide present in vent fluids—were kept together to give rise to primitive living systems. The final step in the process was the synthesis of a lipid membrane that permits the primitive organism to leave the "black smokers" and start an independent life.

Many questions still remain unanswered in the "origin of life on earth" debate. One of those questions is, Did all life begin and evolve at one single place on the planet, or did it begin at several different places? In this context, chemosynthetic environments can be found in oceanic locations other than the ridge venting system— for example, in some continental shelf regions in the presence of methane and hydrogen sulfide. But these regions are the so-called "cold seeps" that lack the high temperature found at the ridge spreading centers.

The prebiotic soup vs. hydrothermal vents controversy is by no means over, and the "soup" hypothesis does not lack supporters. Throughout all the arguments, however, it should be remembered that perhaps the most important question of all has not yet been satisfactorily answered: *What was the first molecule that could replicate itself?*

13

PLATE TECTONICS
AND VOLCANOES

I t was pointed out in Chapter 5 that long before the plate tectonics theory was conceived, it was known that both active and extinct volcanoes are not distributed about the surface of the planet in
a random way, but tend to lie in specific belts, such as the Alpine-
Himalayan and the Circum-Pacific belts. The theory of plate tectonics, however, finally offered an explanation for these volcanic distribution patterns.

The present-day distribution of volcanoes is illustrated in Figure
5-2. It was pointed out in Chapter 5 that this distribution was not
random, but rather volcanic activity was concentrated into specific
belts. In the light of what we know about plate tectonics a new
kind of order can now be imposed on the distribution patterns of
volcanoes. In this new order, three features are outstanding:

▸ At present, there are around 550 active, above-sea-level volcanoes, most of which are located along the margins of adjacent tectonic plates, or over "hot spots" in plate interiors.

▸ More than half of the above-sea-level volcanoes are concentrated in the Circum-Pacific "Ring of Fire."

▸ By far the most important sites of volcanic activity, producing around 75 percent of all lava eruptions, are found below
   sea level at the mid-ocean ridge spreading centers.

Plate tectonics theory recognizes three kinds of volcanic activity: 1) *hot spots*—oceanic or continental intraplate (non-plate-
margin) activity, located over mantle hot spots; 2) *divergent plate*

boundaries—spreading center volcanoes found mainly at the mid-ocean ridges, but also in some continental regions; and 3) *convergent plate boundaries*—subduction zones, including island arc volcanoes.

In general, therefore, volcanic activity can be thought of as occurring in three main environments: hot spots, spreading centers, and subduction zones.

▲

"Hot Spot" Intraplate Volcanic Activity. In the eyes of many people, the typical volcano, such as Mount Etna in Sicily, initially erupts through a central vent and subsequently grows into a cone and perhaps an individual mountain. This is the Hollywood volcano. In reality, however, many volcanic eruptions take place through fissures—that is, long narrow cracks or ruptures—and do not develop into cones. Instead, large volumes of lava flow over the countryside or seabed, sometimes flooding thousands of kilometers of the land and forming vast plateaus termed *flood basalts,* or sometimes lava plateau basalts. In general, flood basalt events have relatively small amounts of pyroclastic ejecta (see Chapter 16).

Flood basalt eruptions have occurred throughout geological time, although fortunately not recently, and have resulted in features such as the Deccan Traps of Central India, the Columbia River Plateau in the western United States, and the Siberian Flats. Oceanic examples include the Kerguelen Plateau in the Indian Ocean, the Agulhas Plateau in the South Atlantic, and Shatsky Rise in the North Pacific.

There is still some dispute over the causes of flood basalts, but it does appear that they are associated with mantle plumes. As the bulbous heads of these plumes of hot rock rise through the mantle toward the surface of the earth, they suffer decompression melting and gather in a pool, which can be thousands of kilometers wide, before erupting.

The Deccan Traps are thought to have originated when seafloor spreading between the Indian and African plates pushed the Indian plate over the deep mantle plume of the Reunion hot spot; in this case, therefore, the plate moved and the heat source remained stationary. However, this model is not universally accepted, and some believe that the Deccan Traps may have been associated with a meteorite impact that resulted in the Shiva Crater, an underwater crater off the west coast of India. The Deccan Traps were formed some 60 million to 70 million years ago at the Cretaceous-Tertiary boundary (the so-called K-T boundary) and originally spewed around 2.5 million cubic kilometers of lava over an area of 1.5 million square kilometers.

The eruption of the Siberian Flats was one of the most important flood basalt eruptions ever to have occurred and is generally thought to have been the largest volcanic eruption in the history of the planet. The flats are a large province, or perhaps an amalgamation of a number of smaller provinces, of igneous rocks formed initially from a mantle plume and injected in multiple layers. The name *flats* derived from the Swedish name for "steps" and refers to the step-like hills found in areas of flood basalts. The volcanic events that occurred in the area of the Siberian Flats lasted for up to a million years at around the time of the Permian-Triassic boundary (the so-called P-Tr boundary), about 250 million years ago. In total, around 3 million to 4 million cubic kilometers of volcanic products covered as much as 2 million square kilometers of ground.

The environmental effects of the Siberian Flats volcanic activity were immense. The explosive nature of the event probably blasted more pyroclastics into the atmosphere than is usual for flood basalts. The immediate area where the eruptions took place would have suffered from lava flows and mud slides. But by far the most harmful feature of the eruptions would have been the emission of volcanic ash and gases, especially sulfur dioxide and carbon dioxide. Ironically, these two gases can have opposing effects on global climate, but operate over different geological timescales. For example, sulfur dioxide yields sulfate aerosols, which, together with volcanic ash and various gases, can spread around the planet and block out sunlight. As a result, over hundreds and perhaps even thousands of years, the surface temperature of the earth dropped and had severe global consequences—even initiating a series of volcanic winters. The climate change could have been sufficient to lead to the extinction of land vegetation and perhaps poison the seas. In contrast, carbon dioxide can result in global warming over millions of years, and it appears that the end of the Permian period is associated with global warming by as much as 6°C.

Flood basalt events of the magnitude of the Deccan Traps or the Siberian Flats are rare, and only eight such events are thought to have occurred during the past 250 million years. Which is fortunate, because flood basalt eruptions have mega-scale environmental impacts, including the most catastrophic disaster of all—*species extinctions*. In fact, a linkage has been proposed between large flood basalt volcanic events and species extinction over geological time. The Permian-Triassic (P-Tr) extinction, which has been called the "Great Dying," occurred approximately 250 million years ago, at the time of the Siberian Flats. It was the most severe extinction in the history of the earth; and it has been estimated that about 95 percent of all marine species and 70 percent of all terrestrial species were lost, and that biological recovery was slow for the next 5 million years. A number of different explanations have been proposed for the P-Tr extinction, and some scientists believe that meteor impacts were involved. In reality, a mixed cause-and-effect sequence may have been at work, with a meteorite impact actually generating flood basalt eruptions, as well as having other environmental effects. The causes of mass extinctions will be the subject of debate for some considerable time, but there can be no doubt that large-scale volcanic activity has played a part in major environmental shocks that have hit the planet in the past.

Volcanic Island Eruptions. Tuzo Wilson became interested in three linear chains of island volcanoes in the Pacific. He found that in each of the chains the islands became younger to the southeast; furthermore, the volcanoes there were still active. By considering each island chain as an entity, Wilson proposed that the older islands were once above a stationary "hot spot," but had moved away as the Pacific plate drifted to the northwest—a hot spot track that offers proof that plate tectonics really do work.

The most famous string of oceanic islands generated by a hot spot is probably the Hawaiian Islands, which formed over 70 million years of volcanic activity as the Pacific plate crossed a stationary hot spot located deep in the mantle. This hot spot partially melts the region below the plate, producing small blobs of magma that eventually erupt as lava. The age of the Hawaiian Islands becomes progressively older toward the northwest, a fact first noticed by James Dwight Dana, who based his thinking on the degree of erosion each island had suffered.

The Hawaiian volcanoes are typical shield volcanoes constructed from many outpourings of lava from a vent or, very often, a number of vents. Two kinds of solidified lava flow are common in the islands: *pahoehoe*, which has a smooth surface, and *aa*, which has a rough, jagged surface with loose debris. Hawaii's volcanoes, which are among the most spectacular in the world, display many of the features most popularly associated with volcanic activity. The eruptions usually begin as lava fountains spouting up to 500 meters high, forming curtains of fire. After a few hours, the eruptions often become localized at a single vent, which leads to the formation of lava streams—the so-called "rivers of fire" that can travel in excess of 50 kilometers an hour down steep slopes. Some of the lava flows may develop a roof, with the flow then being confined below the surface and the lava staying hot longer.

Two volcanoes, Mauna Loa and Kilauea, on the main island are among the most active on earth. Over the past 200 years or so they have erupted, on average, every two or three years. By and large, the dangers posed by the Hawaiian volcanoes are relatively mild, and they have even become tourist attractions. However, volcanic eruptions on Hawaii can be extremely harmful to population centers and agriculture, and in the past there have been examples of entire villages being destroyed.

▲

Spreading Center Volcanic Activity. The intertwined stories of the two great theories of seafloor spreading and plate tectonics pivot around the processes taking place at the spreading centers on the mid-ocean ridge system. The most extensive rift valley on the face of the earth is found along the crest of the mid-ocean ridge, where the new crust is generated, which forces the ocean basins to spread apart. It is important to remember that at the mid-ocean ridge the continental crust is absent, so the basalt rising from below does not undergo compositional modification as it does at the subduction zones. At the ridges, therefore, the products of volcanism remain largely basaltic in composition.

The processes at the mid-ocean ridge take place under the sea, and inevitably much of the evidence on which the theory of seafloor spreading was based relied on arm's-length research, acquired either indirectly (e.g., by remote measurements) or directly (e.g., by viewing from submersibles). The fact remains, however, that almost nowhere can the "furnace of creation" at divergent plate boundaries be witnessed on land. With one exception: *Iceland*.

Iceland lies on the Mid-Atlantic Ridge, a constructive plate boundary where new seafloor is being generated as two plates, in this case the American and the Eurasian plates, are pried apart. Iceland has formed from the eruption of many lava flows, and hot springs, geysers, and geothermal lakes associated with hydrothermal activity abound on the island. Iceland is therefore the place where scientists can study firsthand the creation of new crust.

Harry Hess put his famous paper on seafloor spreading into the scientific literature in 1962, and as if on cue, Iceland provided an example of the kind of volcanism that drives the spreading. The event was the formation of the new island of Surtsey, a volcano on the mid-ocean ridge system where the furnace of creation gives birth to new oceanic crust as the earth responds to the forces of plate tectonics.

Surtsey lies off the southern coast of Iceland, and the emergence of the island was the end-product of a series of eruptions that had started on November 8, 1963, 130 meters below sea level, in the Vestmann Isles submarine volcanic system that is part of the Mid-Atlantic Ridge. Initially, the eruptions had been quenched by the overlying seawater, and the first outward sign of the birth of the island came on the November 14, when a column of black smoke and ash was seen to escape from the sea surface. Then on November 15, a thin ridge of volcanic rock broke the surface of the Atlantic Ocean. During the following days smoke, cinder, ash, and lava bombs were thrown into the atmosphere, and at one point, a column of steam, 6 kilometers high, rose above the island and deposited volcanic ash over the surrounding area. During this period lightning storms, generated by charged particles of ash, and swirling smoke made the nascent island look like a scene from Norse mythology. Fitting, then, that the island was named after Satur, the fire god of local legend.

Over the next few years, the lava and ash began to build up into a cone-shaped island, and Surtsey grew to a maximum size of 3 square kilometers, with an elevation of about 170 meters above sea level. Later, however, the size was reduced due mainly to erosion from wave action and the wind. Surtsey is unlikely to be added to by further eruptions, since the volcano from which it came is thought to be extinct and the island is loosing around 10,000 square meters of its surface every year.

Spreading centers can occur in locations other than mid-ocean ridges, such as within continents. A prime example is the tectonic activity associated with the formation of the East African Rift Zone and the birth of a new ocean at the Red Sea. As spreading develops, probably driven by mantle plumes of hot rock, the crust is stretched and thinned and eventually tensional cracks develop. These cracks are accompanied by faulting, which causes large sections of the crust to sink between parallel faults, generating volcanic activity as magma is forced up through the cracks. In the East African setting, the African continent has started to break up and the Arabian Peninsula has already been torn away from the rest of Africa forming the linear Red Sea, which may develop into the next major ocean.

The Great Rift Valley is a major structure that runs about 6,000 kilometers from Syria in the north to Mozambique in the south, and varies in width from 30–100 kilometers. The rift has been forming for around 30 million years and has produced a vast area of dramatic scenery with lakes, volcanoes, and valleys. It includes features such as Bekaa Valley, the Jordan River, Lake Tiberius (the Sea of Galilee), the below-sea-level Danakil Depression and Dead Sea, and the Red Sea.

Volcanoes associated with the tectonic activity of Africa's Great Rift Valley include the photogenic, snowcapped Mount Kilimanjaro ("Shining Mountain" in Swahili), one of the most famous volcanoes in the world. They also include two volcanoes that are very special because of the nature of the lava they erupt.

Mount Nyiragongo, a stratovolcano with a height of 3,470 meters, is in the Democratic Republic of the Congo. The volcano, which last erupted in 2006, has a lava lake and discharges alkali-rich, very low-silica lava that is extremely fluid in character. The effect is enhanced on Mount Nyiragongo with its steep-sided cone, down which the lava can flow at speeds that can approach almost 100 kilometers per hour.

Mount Ol Doinyo Lengai, "Mountain of God" to the Masai, has a height of 2,890 meters and is situated in Tanzania. It is a classical steep-sided stratovolcano, but is unique because it is the only volcano in the world that yields a *natrocarbonatite* lava. This is a high-alkaline, very low-silica, low-gas, low-temperature lava—around 500°C to 600°C, compared, for example, to 1,100°C for basalt lavas. Natrocarbonatite lavas have more than 50 percent carbonate minerals and less than 10 percent silica. They have an extremely high fluidity, more than any other type of lava, and can flow almost like water, although they look like black oil and have sometimes been mistaken for mudflows. When they solidify they are generally black in color, but turn white on contact with moisture in the air and eventually can become a soft powder.

▲

Subduction Zone Volcanic Activity. Destructive plate margins are often characterized by highly explosive volcanic activity, making them potentially among the most dangerous places on earth in which to live. The volcanoes are generated above slabs of oceanic crust that has been dragged down to depths of 100–150 kilometers during subduction. More than half of the total of the world's active above-sea-level volcanoes are concentrated in the seismically active Circum-Pacific Ring of Fire, a 40,000 kilometer horseshoe-shaped zone that partially encloses the margins of the Pacific Ocean and contains around 75 percent of all the planet's volcanoes that erupt above sea level (see Figure 13-1).

The Ring of Fire contains several distinctive features. These include: *ocean trenches* (e.g., the Peru-Chile Trench, the Marianas Trench, the Aleutian Trench); *island arcs,* which are strings of volcanic islands parallel to the trenches on their landward sides (e.g., the Kuril Islands, the Aleutian Islands, Java, New Guinea); and *mountain chains,* or cordillera, formed in continental crust above subduction zones (e.g., the Andes, the Cascade Range, and the Rockies). The common factor in all these features is subduction, and most of the Ring of Fire volcanoes are the result of convergent plate subduction activity.

In the Ring of Fire, the relatively thin oceanic crust of the Pacific Plate is thrust beneath a continental crust plate. The resultant subduction gives rise to explosive eruptions. In fact, subduction is re-

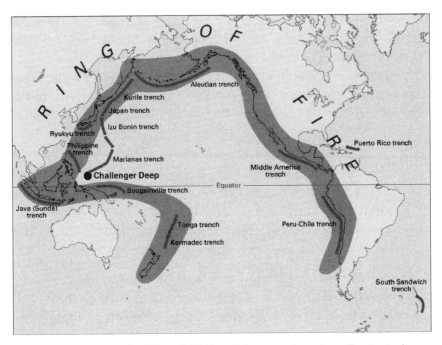

Figure 13-1. The Pacific "Ring of Fire" and deep-sea trenches. The tectonic activity is the result of old seafloor being subducted into the mantle via the deep-sea trenches that define the shape of the ring. The trenches are formed at plate boundaries where oceanic crust (e.g., the Pacific Plate) is thrust beneath continental plate, generating intense volcanic and earthquake activity. Credit: United States Geological Survey.

sponsible for a large proportion of the total geological violence manifested in the form of volcanic and earthquake activity that is visited on the planet. The reason that subduction volcanoes are explosive is complex, but the principal underlying factor is that the magma generated in the process contains a relatively high proportion of gases, such as steam—important because water plays an extremely significant role in both the melting of the mantle and volcanic eruptions. The subducted oceanic crust carries a sediment cover, and as it is dragged down into the mantle it becomes heated and eventually dehydrates. As it does so, water is released into the mantle, causing it to melt at a lower than normal temperature. The melted material, plus the contained water, then explodes as the steam, and other gases, are released violently. As a result, volcanic activity at subduction zones tends to be more violent than at locations where oceanic crust does not get dragged down into the mantle—such as at the spreading centers of the mid-ocean ridge system or above-mantle hot spots.

Central to everything in the Ring of Fire is the large Pacific Plate. This is a relatively thin oceanic plate, and volcanic activity in the Ring of Fire can be related to the way it collides, and juggles for position, with a series of other plates:

- ▸ The tectonic activity along the eastern rim of the ring is caused by the Nazca and Cocos plates being subducted under the South American Plate, with the associated features of the Peru-Chile Trench and landward, the Andes Mountains.

- ▸ Northward, the Pacific and Juan de Fuca plates are being subducted under the North American Plate, with the associated features of the Middle America Trench and landward of this, the Rocky Mountains.

- ▸ Thus, there is a spine of mountains, or cordillera, running through South, Central, and North America that makes up the eastern half of the Ring of Fire.

- ▸ Further north, the volcanic activity in Canada and Alaska is the result not of subduction, but of the Pacific Plate sliding past the Queen Charlotte Fault and cracking the North American Plate.

- ▸ On its northern boundary, the Pacific Plate is subducted below the Aleutian Trench.

- ▸ Along the western rim of the Ring of Fire horseshoe, the situation is more confused. Here, the Pacific Plate undergoes subduction below the Eurasian Plate and the Indo-Australian Plate at a number of oceanic trenches (e.g., Kuril, Japan, Ryukyu, Izu-Bonin, and Philippine) and associated island arcs.

- ▸ At the top of the rim lie the intensively active volcanic regions of the Kamchatka Peninsula and parts of China and Japan.

▸ Further toward the equator, the Philippine Plate is being subducted under the Eurasian Plate, at the Marianas Trench, which contains the Challenger Deep (the deepest part of the ocean) and the Philippine Trench.

▸ Along the southwestern edge of the horseshoe, the Pacific Plate is being subducted beneath the Indo-Australian Plate at the Java and Bougainville trenches, and further south at the Tonga and Kermadec trenches.

▸ New Zealand, which is one of the most seismically active regions in the world, lies at the extremity of the Ring of Fire on the boundary of the Pacific and Australian plates. The country has a high density of active volcanoes that occur in six areas, five in the North Island and one offshore. White Island, Ruapehu, and Ngauruhoe are three of the most active volcanoes in the region.

▸ Moving around the Pacific Ring of Fire, the important volcanoes include Mount Fuji, Mount Erebus, Krakatau, Mount Pinatubo, White Island, Popocatepetl, and Mount St. Helens.

▲

The Mediterranean Sea is almost landlocked, being bordered by Europe to the north, Africa to the south, Asia to the east, and the narrow Straits of Gibraltar in the west. The best estimate is that the Mediterranean was formed during rifting of the African and Eurasian plates in the Late Triassic and Early Jurassic periods. Although the tectonic history of the central and eastern Mediterranean region is extremely complex, much of it revolves around the collisions and recessions of the African and Eurasian tectonic plates, sometimes via a series of miniplates. The key process is the subduction of the African Plate beneath the Eurasian Plate, but the complexity lies in the detail that involves an interplay between subduction and collision boundaries at a number of arcs.

The *Calabrian Arc* runs down the Italian mainland into Sicily and toward Tunisia, and it is the site of the subduction on the African-Eurasian plate boundary.

The *Hellenic Arc* is the most active seismic zone between the African and Eurasian plates, and extends from the western Peloponnesian islands, through Crete and Rhodes into Western Turkey. It is mainly a subduction site.

The *Cyprian Arc* runs from Cyprus onto the mainland of Turkey. It is thought that the northwestern part of the arc is a subduction boundary, but that the southeast section is a collision boundary.

The collisions between plates at locations like these generate volcanism, earthquakes, and mountain building—most notably, the Alps.

The *Antilles Arc* is a chain of volcanic islands in the eastern Caribbean extending for more than 800 kilometers from the Virgin Islands in the north to the coast of Venezuela in the south. The arc was formed in the zone where the Atlantic seafloor is subducted under the western edge of the Caribbean Plate, a process known locally as *soufriere,* which is the source of "sulfur."

Some extremely vigorous volcanic eruptions have occurred on the Antilles Arc, bringing with them destruction and death—the most disastrous being the 1902 eruption of Mount Pelée on the island of Martinique (see Panel 7). The reason for this destructive activity is that the magma involved is low in silica, and as a result the eruptions are highly explosive. In addition, they often involve dangerous pyroclastic flows, such as those formed during the eruption that began in 1995 in the Soufriere Hills on the island of Montserrat. The eruption flung a cloud of superheated ash and gases five miles into the air and resulted in the destruction of the capital, Plymouth.

Clearly, then, the distributions of volcanoes, and the types of volcanic activity, can be adequately explained by the theory of plate tectonics.

14
PLATE TECTONICS
AND EARTHQUAKES

———— 🌍 ————

Although earthquakes can be triggered by a variety of mecha-nisms, such as volcanoes, meteorite impacts, and the explosion of nuclear devices, most *natural* earthquakes are caused by tectonic factors and are bound up in the way in which the earth continually adjusts to the great strains placed upon it—by releasing energy through faulting.

Faults are rock fractures causing displacements between adjacent segments of the earth's crust. Originally, many geoscientists believed that earthquakes generated faults. Then, as we have already seen, Fielding Reid came up with the elastic rebound theory, in which he proposed that earthquakes were initiated by faults, and not the other way around.

In fact, most earthquakes are the result of the elastic rebound of previously stored energy.

Earthquakes tend to recur along the fault lines, which is not surprising since they are lines of weakness in the crust. Faults are formed when the stress that is built up, as rocks try to move relative to each other, exceeds the strain threshold and releases strain energy. The two blocks that move in a fault are termed the *hanging wall*, which is above the fault, and the *footwall*, which is below the fault. Slip, or slippage, is the movement on the fault plane—the plane along which the relative motion takes place. The way in which the slip, or relative motion, occurs characterizes a fault.

There are several different kinds of faults, but we need only consider three (normal, reverse, and transform) that occur at plate boundaries, and that are dependent on the way blocks of crust (plates) move relative to each other. In this way, the faults can be categorized according to whether the movement involved is vertical (dip-slip faults) or horizontal (strike-slip faults). These various faults are illustrated in Figure 14-1.

Dip-Slip Faults. Here, the movement is essentially vertical and tensional stress causes the crust to break across a slope. A *normal fault* occurs when the stress is caused by extending or stretching the crust, and it is sometimes called an extensional, or gravity, fault. It is formed when tension forces the hanging wall to slide down the fault plane, relative to the footwall, under the influence of gravity. A *reverse fault* occurs when stress is caused by compression of the crust, which forces the hanging wall to move upward relative to the footwall. As a result, the crust is shortened.

Strike-Slip, or Transform, Faults. They are formed when blocks of crust slide past each other horizontally with very little vertical motion. These faults are characteristic of the mid-ocean ridges.

Earthquakes are intimately associated with the relationship between plate tectonics and faulting, and most, but certainly not all, earthquakes occur at plate boundaries.

Horizontal displacements between adjacent segments of the earth's crust may be on a scale of millimeters to several kilometers. Individual plate boundaries, or margins, are subjected to different interplate stresses that, in turn, are associated with different types of faults. As a result, various kinds of earthquakes are generated. And by linking together the type of plate margin, the type of force applied to a rock, and the type of fault, earthquakes can be classified in terms of their tectonic setting.

Spreading Center Earthquakes. Here, existing crust is forced apart as new crust is formed (creation). Normal faulting predominates at these boundaries; for example, in an oceanic setting at the mid-ocean ridge spreading centers, and in a continental setting at the Great Rift Valley of East Africa. The earthquakes generated by these normal faults are shallow, usually less than 30 kilometers, and tend to be less than magnitude 8 on the Richter scale.

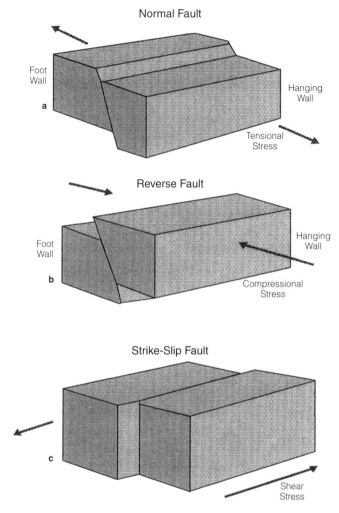

Figure 14-1. Earthquakes and the main fault types that occur at plate boundaries. Normal and reverse faults are dip-slip types and involve vertical movement. Transform (or strike-slip) faults involve horizontal movement. The arrows indicate stresses that cause the fault motion.

a. A *normal fault* occurs when the stress is caused by extending, or stretching, the crust, and it is sometimes called an extensional or gravity fault. It is formed when tension forces the rocks above the fault plane (the hanging wall) to slide down, relative to the other side of the fault (the footwall).

b. A *reverse fault* occurs when stress is caused by compression of the crust, which forces the hanging wall to move upward relative to the footwall. As a result, the crust is shortened.

c. A *transform, or strike-slip, fault* is formed when blocks of crust slide past each other horizontally with relatively little vertical motion. Credit: Stefanie J. Baxter, Delaware Geological Survey.

Subduction Zone Earthquakes. Here, one crustal plate is forced under another (destruction). Reverse or thrust faults are characteristic of such a plate margin. Earthquakes in this tectonic setting can have depths ranging from very shallow to several hundreds of kilometers deep and often lie in Benioff zones (see Chapter 10) as the subducting plate continues to undergo brittle fracture as it is forced down below the plate above. Benioff zones dip downward at approximately 40–60 degrees from a deep-sea trench, following the subducting oceanic plate as it dives several hundred kilometers into the mantle under the continental plate above. This kind of earthquake distribution has played a large part in our understanding of the process of subduction, and plotting the focal points of earthquakes occurring under the crust has allowed scientists to "see" how one plate dips beneath another. Subduction boundaries are the home of some of the world's most destructive earthquakes that can reach magnitude 9 on the Richter scale.

Transform Fault Earthquakes. These are fracture zones where there is neither creation nor destruction of crust, but where segments of crust slide horizontally past each other—especially at the mid-ocean ridge. Here, earthquakes tend to be shallow and of low magnitude, but under some conditions transform fault earthquakes can cause considerable damage. The most famous of all transform faults, the San Andreas Fault, has wreaked much of its earthquake damage inland and in one earthquake event devastated San Francisco (see Panel 8). Another well-known transform fault is the North Anatolian Fault in Turkey, which is one of the most seismically active regions in the world.

Intraplate Earthquakes. This category includes less than about 10 percent of the world's earthquakes, and although they are not at plate margins at present, their origin may be related to ancient margin faults that are now in the interior of tectonic plates. One of the most important locations of intraplate earthquakes is found in Eastern Asia, where earthquakes can occur more than a thousand kilometers from a plate boundary.

No one really knows how many earthquakes occur over a given period of time, one reason being that many earthquakes are not recorded. The United States Geological Survey has estimated that more than 3 million earthquakes occur every year, but the vast majority of them are far too weak to cause harm in any way. Even so, each year around 150,000 earthquakes worldwide have effects that can be picked up by humans. But the worst damage is obviously the result of large earthquakes hitting highly populated regions.

The global distribution of earthquakes is illustrated in Figure 5-2, and it was pointed out in Chapter 5 that like volcanoes, earthquakes are not distributed randomly around the earth. In fact, around 95 percent of all earthquakes tend to occur in a series of zones, or seismic belts. Furthermore, the distribution of earthquakes has a number of striking similarities to that of volcanoes (see Chapter 5, Figures 5-2 and 5-3). At this point, it is useful to summarize again this earthquake distribution. Worldwide, there are three predominant seismic belts:

▸ The Circum-Pacific Ring of Fire belt accounts for almost 80 percent of the seismic energy released on the planet.

▸ The Mediterranean-Asiatic, or Alpide, belt—which stretches from the islands of Java, Bali, and Timor, through the Himalayas, the Mediterranean, and out into the Atlantic—accounts for around 15 percent of the world's earthquakes.

▸ The Mid-Ocean Ridge belt is found at the mid-ocean ridge spreading centers. Earthquakes here tend to have a magnitude of less than 5.

As was the case for volcanoes, a new order can now be imposed on the distribution patterns of the earthquakes in the light of plate tectonics. The major seismic belts occur along the margins of what we now know to be tectonic plates—especially at the convergent and transform fault margins. Only the intraplate earthquakes are not concentrated into specific bands, but include some of the biggest seismic events ever recorded. It is apparent, then, that the generation of many earthquakes can be described within the setting of plate tectonics.

15
PLATE TECTONICS
AND TSUNAMIS

───────── 🜨 ─────────

The generation of tsunamis is inextricably bound up with the earthquakes that provide triggers for most of these killer waves. We have already seen that many of the world's earthquakes occur at plate boundaries where the plates jostle with each other, which we explored at length in the last chapter.

Of the earthquakes associated with the three kinds of plate boundary—transform fault, divergent, and convergent—it is the latter, the convergent subduction boundaries, that are especially linked to tsunamis.

Tsunamis, which were first reported in 2000 BC off the Syrian coast, can be generated in all the major oceans, and indeed in any large body of water, but they appear to have occurred most frequently in the Pacific Ocean and the marginal seas surrounding it. Here, tsunami generation is closely related to earthquakes and, to a lesser extent, the volcanoes of the subduction zone associated with the Pacific Ring of Fire. Japan has the unenviable record of having the highest number of recorded tsunamis in the world, averaging almost one every seven years. Some of them have been extremely hazardous, destroying complete coastal towns. Alaska, which is part of the Ring of Fire, and Hawaii (a "hot spot" region) are also in the forefront of the tsunami league.

The Atlantic Ocean has many fewer subduction zones than the Pacific, and those that are located there (e.g., around the Caribbean) are considerably smaller and less seismically active. As a result, the Atlantic has only a few percent of all recorded tsunamis.

Nonetheless, tsunamis have occurred in this ocean, the most destructive being the one that followed the Lisbon earthquake in 1755 (see Panel 5); this tsunami wiped out the city and had a death toll of around 60,000. Other Atlantic tsunamis have had a much smaller loss of life, ranging from a few hundred to a few thousand people.

Although they are not as frequent as in the Pacific Ocean, the tsunamis that occur in the Indian Ocean can pose extreme hazards to coastal populations, because many of them inhabit low-lying ground. The Indian plate is active on both its east and west margins, and the Indian Ocean has a number of subduction sites. On the eastern boundary of the Indian subcontinent, the Australian and Eurasian plates collide with the Indian plate, and on the western boundary off the Makran coast of Pakistan and the west coast of India, the Arabian and Iranian plates collide with the Indian plate. Further east, the subduction of the Indo-Australian plate beneath the Burma and Sunda plates, around the Sunda Trench, is a very active earthquake and volcanic region that has seen the generation of two of the most destructive tsunamis in history. The first of these events, in 1883 came after the Krakatau volcanic eruption (see Panel 3), and the overall loss of life was around 35,000. The second, in 2004, followed an earthquake off the west coast of Northern Sumatra. The tsunami that followed devastated coastal areas in many countries and was responsible for around 225,000 deaths, the greatest ever known loss of life in a tsunami (see Panel 15). Most of the remaining tsunamis recorded in the Indian Ocean have had relatively small, but nonetheless extremely significant, death tolls, ranging between a few hundred and several tens of thousands.

The Eastern Mediterranean is one of the most seismically active regions of the world. Earthquakes are generated along the North Anatolian Fault where the Anatolian plate grinds against the African plate, and the Mediterranean Sea has been subjected to a number of large tsunamis in the past. The most destructive hit the Greek islands between 1700–1400 BC. A tsunami that occurred in 1410 BC, for example, killed more than 100,000 people, on the basis of modern estimates, and there is historical evidence that tsunamis caused extensive damage to the Minoan civilization. In addition to earthquakes, several volcanoes in the region, including Etna, Stromboli, and Vesuvius, have had eruptions that resulted in tsunamis. One of the most destructive earthquake-induced tsunamis in the Mediterranean Sea stuck the Messina Strait, between Italy and Sicily, in 1908 (see Panel 9). This was the result of a magnitude 7.5 earthquake, and more than 100,000 people were killed. In terms of loss of life, however, most other tsunamis in the Mediterranean have been relatively small in scale, with death tolls in the range of less than ten to around 1,500. Seas adjacent to the Mediterranean can also be prone to tsunamis. The Sea of Marmara, which lies in an extremely seismically active region associated with the North Anatolian Fault, is a tsunami-prone region.

On average, then, in the tsunami "devastation league" at the present time the Pacific comes out top, followed by the Indian Ocean and then the Mediterranean, with the Atlantic a good way behind. It should not be forgotten, however, that the most destructive of all recorded tsunamis occurred in the Indian Ocean in 2004.

16
PREDICTING AND MITIGATING AGAINST VOLCANOES, EARTHQUAKES, AND TSUNAMIS

The extent to which the prediction and monitoring of any natural phenomenon can be successful depends ultimately on how well we understand the science behind that phenomenon. We now have a theory, plate tectonics, that can satisfactorily explain the formation of volcanoes, earthquakes, and tsunamis.

So we know why they occur. But how close are we to being able to use this scientific understanding to predict the occurrences of the "trinity of natural disasters," and how well can we mitigate against the damage they bring?

PANEL 12
RISKS AND HAZARDS

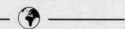

Before attempting to assess the harmful effects that volcanoes, earthquakes, and tsunamis have on the environment, it is useful to define what we mean by risks and hazards. There are a number of definitions of these terms, some of which are often confusing because they have been devised for different reasons. Without becoming embroiled in legal jargon, however, it is possible to lay down a few guidelines.

In our definition, a *hazard* offers a potential threat to human life, property, the environment, or to all three. The threats are delivered by *hazard events* that, in this context, are associated with natural phenomena: that is, volcanoes, earthquakes, and tsunamis. *Risk* refers to the probability of a potential hazard causing harm.

It is important that society manages both potential hazards, and hazard events, in a manner that restricts the damage they cause to life, property, and the environment. To do so, it is necessary to have a management strategy in place. Such a strategy involves a complex interplay between scientists, local and national governments, and especially in the Third World, aid agencies. The overall aims of such a strategy are to minimize loss of life and damage to property, to preserve the infrastructure of society, to offer assistance as rapidly as possible to disaster victims, and to aid recovery. To this end, the strategy should involve 1) a predisaster phase, involving monitoring, prediction, public education, and mitigation; 2) a disaster phase, involving relief preparedness; 3) a postdisaster phase, involving response, recovery, and reconstruction.

Today, these aims are often incorporated in some form of a *Disaster Management Cycle,* which typically consists of a number of elements.

Prediction. To provide **warning** of an imminent disaster before it happens, it is necessary, when possible, to put monitoring techniques in place. The mantra for any form of disaster prediction is, "Where, when, and how powerful?"

Mitigation. Methods for **minimizing** the effects of a disaster before it happens may include, for example, introducing building codes that incorporate specific safety measures.

Preparedness and Response. To be able to promptly react to a disaster **after** it has happened, the requirements are putting plans of action in place and setting up a strategy for their implementation. Such plans usually involve multi-agency cooperation, the coordination and training of emergency services via exercises, the provision of population warning systems, and the establishment of vital amenities such as water supply, emergency shelters, and basic medical care.

Recovery. This phase involves the restoration of an area, or a community, to the predisaster state; for example, by instigating rebuilding projects and carrying out infrastructure repair.

This general type of approach may be expressed as:

Predictions ➡ Precautions ➡ Recovery

A framework of this kind allows a community to assess how prepared it is to handle natural disasters.

For our purposes, it is useful to describe the disasters arising from volcanoes, earthquakes, and tsunamis within a common framework that links 1) the fundamental science of their origins, 2) the potential hazards they pose to the environment, 3) the warning systems that can be used to predict them, and 4) the steps that can be taken to mitigate against the damage they might cause.

It is important, too, to stress that the role the public can play before, during, and after disasters must never be underestimated. In this respect, *public education* is an extremely important part of any approach to hazard management, if only because of the need to have the public on board. Awareness of potential environmental hazards can be part of a formal educational approach at all levels, from primary schools to universities, or it can be channeled through the media and the Internet.

Many institutions disseminate basic science and hazard information via the Internet. An excellent example of how to approach this kind of public education is provided by the programs run by the United States Geological Survey (USGS) and the British Geological Survey (BSG). The programs designed by both of these bodies include a strong Internet component that presents information on the basic science behind volcanoes, earthquakes, and tsunamis, as well as the hazards associated with them. Furthermore, in addition to a purely educational function, the USGS especially puts out hazard information on a global scale. Other programs like those of the USGS and the BGS are in place in various countries around the world.

The definitions of hazards and risks we will use are described in Panel 12. The principle volcanic hazards arise from material spewed out into the environment. From earthquakes, the hazards come from disturbances to the ground, and from tsunamis they originate from great waves.

Thus, each of the three disasters brings its own particular dangers.

VOLCANOES: "FIRE AND MOLTEN ROCK"

The material thrown out during a volcanic eruption consists of lava, tephra (including ash), and gases—all of which can generate extreme environmental hazards.

Lava flows are most common in effusive, rather than explosive, volcanic events. In the immediate locality of an eruption hot lava flows, which are sheets of molten rock, can travel for several kilometers, and speeds of movement vary from very fast catastrophic rushes to gentle rolling movements. Many lava flows are restricted to the immediate vicinity of a volcano, but some eruptions, such as flood basalts, can cover very large areas. Because of their high temperature, lava flows can wreck havoc on anything in their path, either by setting it alight or by engulfing it in what will cool to become solid rock.

The solid material, or ejecta, thrown into the air by a volcano is termed *tephra*, or *pyroclastics*. This material includes blocks and bombs of various kinds, lapilli (scoria, pumice, and cinders), glass shards, and ash and dust. Volcanic ash can cover a wide area and be extremely dangerous when deposited in thick blankets, as happened during the eruption of Vesuvius (see Panel 1). Even thin ash layers can disrupt living conditions—for example, by clogging up drainage systems and affecting crops.

One of the most dangerous of all "along the ground" volcanic threats comes from pyroclastic flows, also called *nuée ardente* ("glowing cloud"), which can cause death by burning or suffocation. Pyroclastic flows are dense clouds consisting of gases and small particles of lava that flow down the slopes of volcanoes at very high speeds, leaving devastation in their wake. When a high proportion of gas is present, the flows are termed *pyroclastic surges*. Temperatures of pyroclastic flows are high, in the range 500°C to 1,000°C, and the flows travel at speeds usually between 50–100 kilometers per hour along the ground. This combination of high temperature and high speed makes pyroclastic flows particularly dangerous, and they can be transported for long distances. They can even move over water surfaces, as was evident during the eruption of Mount Pelée (see Panel 7).

Mudflows, sometimes called *lahars* or debris flows, are gravity-driven mixtures of mud and water that sweep downhill. They can be up to hundreds of meters wide, tens of meters deep, and travel at speeds of as much as 900 kilometers per hour. Mudflows have caused extensive fatalities in the past—more than 23,000 people died when mudflows from the 1985 eruption of the Nevado del Ruiz volcano in Colombia hit the town of Armero. Other types of gravity flows include volcanic landslides and flash floods, and all of these flows can have devastating effects on crops and habitats.

Volcanic gases that can rise to tens of kilometers into the air as part of volcanic plumes are dominated by steam (water vapor), which usually makes up between 60 percent and 90 percent of the gaseous components, and carbon dioxide. In addition, lesser amounts of other gases, such as hydrogen, are present in volcanic plumes. Some of these gases are toxic; for example, sulfur dioxide can combine with water vapor in the air to form acidic aerosols or create volcanic smog by reacting with sunlight, oxygen, dust, and water.

Volcanoes can also generate a number of secondary hazards. Flooding can occur when drainage systems become blocked by volcanic debris, sometimes forming a dam that might break with catastrophic consequences. Famine can be another secondary effect of volcanic activity, arising from crop failure due to extensive ash deposits blanketing the crops. Famine was the major cause of death following the devastating Tambora volcanic eruption in 1815 (see Panel 2).

Volcanic activity can have a variety of effects on weather and climate. They range from volcanic dust and ash particles erupted into the higher atmosphere, leading to dramatically colored sunsets, to the potentially very dangerous lowering of global temperatures for up to several years after a major eruption. This cooling can have very severe consequences for humans, a prime example being the "year without a summer" that followed the eruption of Mount Tambora in 1815. The decrease in temperature following a volcanic event is caused by dust particles in the air or, more dangerously, by sulfur dioxide forming sulfuric acid droplets that are very efficient at reflecting sunlight and forming sulfur hazes that cut down the radiation that reaches the surface of the earth. In addition to Tambora, climatic effects have been found to occur after several major eruptions—including Krakatau (1893) and Mount Pinatubo (1991).

It has been suggested that volcanoes have played a role in setting the composition of the atmosphere by acting as giant reactors. Submarine volcanoes are more reducing than terrestrial volcanoes and are thought to have scrubbed oxygen from the atmosphere, perhaps by combining with gases in a volcanic plume to remove the oxygen. When the ratio between submarine and territorial volcanoes shifted toward the terrestrial types, which it apparently did around 2.5 billion years ago, more oxygen was allowed to accumulate in the atmosphere. Much of this oxygen was produced by bacteria during photosynthesis, a process that releases oxygen. This shift to an oxygen-rich atmosphere had tremendous consequences for the way in which life developed, and defined the way the planet became populated.

Volcanoes have also been cited as savers of the planet during the so-called "snowball earth" episode, which is thought by some to have occurred 600 million to 800 million years ago when the earth was almost completely covered by ice. One explanation given is that enhanced weathering of rocks led to reactions that absorbed carbon dioxide (CO_2), a "greenhouse" gas, from the atmosphere, leading to such a decrease in global temperatures that ice took over the entire planet. An injection of large quantities of CO_2 into the atmosphere was required to overcome the "snowball earth" effect, and one suggestion is that volcanoes projecting above the ice cover would have put the greenhouse gas into the air. That the "snowball earth" phenomenon ever occurred is strongly contested by a number of scientists, but if indeed the earth was at one time completely covered with ice, it must have had a profound effect on life and evolution.

Clearly, volcanoes provide a raft of hazards that play out on a global, as well as a local, scale. Table 16-1 lists the most destructive volcanic eruptions ever known to have occurred.

An *active* volcano can be defined simply as one that has erupted within recorded history. By the same token, a volcano that has not erupted during recorded history is said to be *extinct*. A *dormant* volcano lies somewhere between the two. Volcanoes, together with earthquakes, are not distributed at random around the earth, but tend to be concentrated in plate tectonic–related zones. Because of this, the extent to which humankind is at most risk will also be concentrated in these hazard zones, although some of the dangers can be worldwide. For example, when great plumes of volcanic gases, dust, and ash are thrown into the air they can be transported for large distances, sometimes around the world.

Techniques for predicting and monitoring potential volcanic hazards range from simply keeping an eye on a known danger spot, to employing the most sophisticated satellite technology:

▸ *Geological History*. Historical data can provide a long-term footprint that can be a useful indication of the chances of a volcano erupting again.

▸ *Magma Movement*. In the light of current knowledge, perhaps the most essential parameter that indicates potential volcanic activity is the movement of magma into a magma chamber, and its subsequent buildup beneath a volcano.

▸ *Earthquake Monitoring Networks*. Earthquakes of various intensities, often starting in small-scale seismic swarms, are a prelude to many volcanic eruptions and provide one of the earliest indicators that a volcano may be stirring.

▸ *Satellite Data*. From as early as 1964, when NASA first put

TABLE 16-1
The Ten Most Destructive Volcanic Eruptions in Recorded History

	Location	Date	Fatalities*	Principal cause of death
Mt. Tambora	Indonesia	1815	92,000	Famine
Krakatau	Indonesia	1883	36,000	Tsunami
Mt. Pelée	Martinique, Antilles Island Arc	1902	29,000	Pyroclastic flows
Ruiz	Colombia, Pacific Ring of Fire	1985	23,000	Mudflows
Mt. Unzen	Japan, Pacific Ring of Fire	1792	14,000	Tsunami
Mt. Vesuvius	Italy	AD 79	3,500–20,000	Ash falls
Laki	Iceland	1784	9,000	Famine
Kelut	Indonesia	1919	5,000	Mudflows
Galunggung	Indonesia	1882	4,000	Mudflows
Mt. Vesuvius	Italy	1631	3,500	Ash falls

The data used in compiling this table is taken from a variety of sources.
* Approximate death tolls.

its High Resolution Infrared Radiometer (HRIR) into the air, volcanic data has been gathered from satellites. Sophisticated current techniques include NASA's Total Ozone Mapping Spectrometer (TOMS), which provides information on the amount of sulfur dioxide (SO_2) released in volcanic eruptions. In addition, the movement of material, such as ash, ejected in volcanic eruptions has been monitored by satellite instrumentation. NASA can now also detect small changes in topography that can be useful in volcanic predictions, using satellite-mounted Synthetic Aperture Radar (SAR). Other nations also have similar satellite programs, such as the Japanese-run Advanced Earth Observation System (ADEOS).

▸ *Hazard Maps.* They are an important tool in predicting the effects of volcanic activity by identifying the geographic areas that would be affected by an eruption. Ideally, such maps would delineate areas likely to suffer from dangers such as pyroclastic flows and mudflows. The maps can be used in various aspects of volcanic disaster planning, including the identification of evacuation routes.

These techniques, in conjunction with a variety of other approaches—including electrical resistivity, ground deformation, S-wave, heat flow, gas analysis, and hydrologic networks—are used to predict, and subsequently monitor the effect of volcanic activity.

In any response to a potential disaster, the widest possible dissemination of information to the public is critically important. In 1974, the U.S. Congress made the United States Geological Survey (USGS) the lead federal agency for the provision of volcanic hazard warnings. The USGS Volcano Hazards Program, which has a Volcanic Disaster Assistance Program (VDAP) for work overseas attached to it, is designed to carry out a number of functions. These roles include monitoring volcanoes, producing weekly updates on worldwide volcanic activity, providing volcanic advance warning systems, and offering help with the emergency planning needed to tackle volcanic disasters.

The VDAP dispatches a rapid-response "volcano crisis" team anywhere in the world and has responded to more than ten potential volcanic disasters in Africa, Asia, the South Pacific, the Caribbean, and Central and South America. These disasters included the 1991 eruption of Mount Pinatubo, in which more than 75,000 people were safely evacuated before the eruption took place. VDAP works closely with local professionals and is involved in training programs for both scientists and public officials from overseas countries. The kind of assistance provided includes the rapid deployment of a mobile volcano observatory and the long-term installation of permanent sensors.

The USGS operates volcanic observatories, in cooperation with local authorities, in Hawaii and Alaska. Both of these observatories have pioneered many of the techniques now used in monitoring volcanic activity, especially the Hawaiian facility. Following the Mount St. Helens eruption, the USGS set up a third observatory, the Cascades Volcanic Observatory, in Vancouver, charged with monitoring Mount St. Helens and other volcanoes in the Cascade Range.

Volcanic monitoring and disaster planning are costly exercises that require bold decision making for their implementation. As population centers move closer to volcanic sites, the monitoring problem remains acute—according to data supplied by VDAP, at present less than 5 percent of historically active volcanoes are adequately monitored, and less than a third are monitored at all.

Mount St. Helens offers an example of how both volcanic monitoring and volcanic hazard planning operate in the modern world. The volcano lies in the Cascade Range, which is part of the Pacific "Ring of Fire." It is a stratovolcano that achieved lasting fame when it underwent a catastrophic eruption on May 18, 1980. This eruption, which was the most damaging in U.S. history, was responsible for the greatest volcanic landslide in recent times and blasted a column of ash in the air to a height of 20 kilometers. Moving at up to 100 kilometers an hour, a massive debris flow, 2.5 cubic kilometers in volume, blocked rivers and destroyed trees. Fifty-seven people died, and fish and wildlife suffered losses. In addition, homes were destroyed, bridges wrecked, and sections of highways and railways put out of action. The loss of life, however, was not as high as it would have been if a hazard warning had not been posted and a "restricted access" zone not been established.

The monitoring of the Mount St. Helens site in 1980 involved a number of elements. There was round-the-clock monitoring of seismic activity, and as the frequency of earthquakes increased, discussions among scientists and local officials led to the closing off of the upper slopes of the volcano because of avalanche risks. An emergency coordination center was set up, and on March 25 the restricted area was extended. On March 27, a Hazard Watch was put in place, and on that day the volcano erupted for the first time in over a hundred years. The eruptions continued over the next few weeks and other restrictions were brought in, with the United States Forest Service (USFS) playing the lead role. On April 1, a volcanic hazards map was drawn up and a news release made the public aware of the increasing risk of danger. Then on April 9, a contingency plan was implemented.

While all this was happening, a bulge was developing on the north flank of the volcano, which, if it had collapsed, could have been a trigger for a major eruption. As the situation developed there were no changes that could unequivocally indicate onset of the actual eruption, but nevertheless the closure zones around the volcano were enforced—and without doubt these measures saved many lives when Mount St. Helens finally blew its top in the cataclysmic eruption of May 18, 1980.

Since the eruption, increasingly sophisticated state-of-the-art monitoring is being carried out on Mount St. Helens, and the volcano is now being watched by a series of civil and military agencies that include the USGS Cascades Volcano Observatory. As a result, a revised Mount St. Helens Volcano Response Plan was brought out in 2006. At its core, the aim of the plan is to provide a coordinated response from the cooperating agencies in order to manage a raft of volcanic activities ranging from small isolated occurrences to multiple events that would require an incident commander and/or a unified area command.

EARTHQUAKES: "WHEN THE EARTH MOVES"

Earthquakes are insidious dangers that lurk out of sight beneath the surface of the earth and wreck their damage by movements of the ground that last, at most, for no more than a few minutes. But in that short time, earthquakes generate a series of major hazards.

In strong earthquakes, violent earth shaking can occur and may be accompanied by rupture of the ground and its permanent displacement along fault lines, such as the San Andreas Fault in California. One constraint, wherever an earthquake is located, is local geological conditions; ground underlain by solid rock will be less affected than ground underlain by unconsolidated sediment, for example. The effect that ground shaking has on the environment also depends on the location of an event, as well as its strength. An earthquake of magnitude 9.0 might have little effect on people if the epicenter was in a particularly remote region, and, in fact, most earthquakes do occur in remote, sparsely populated areas. When they happen in highly populated areas, however, damage from surface waves, which have a low frequency and large amplitude, can be catastrophic, causing the ground to undulate rapidly. In this context, rapid ground shaking of even an inch or two would be sufficient to damage most buildings not specially strengthened for earthquake resistance. As well as affecting structures, ground shaking can buckle roads and rail tracks.

Ground shaking may be regarded as *primary* earthquake damage and can result in considerable loss of life. But some of the *secondary* effects resulting from earthquakes can be equally savage. Landslides and avalanches have the potential to be particularly damaging. In 1970, an earthquake with a magnitude of 7.9, centered under the Pacific Ocean about 30 kilometers from the coast, struck Peru and, among other effects, triggered a massive landslide/avalanche on Mount Huascaran. The slide, which consisted of a 900-meter wide torrent of ice, rock, and water, traveled at speeds of up to 200 kilometers per hour and destroyed the towns of Ranrahirca and Yungay. The final toll of the earthquake was between 70,000 and 80,000 dead.

Soil liquefaction can be an important hazard when ground shaking caused by an earthquake mixes sand, soil, and groundwater together into a semiliquid state, so that the ground becomes soft and unstable. This can result in buildings canting at an angle or sinking into the underlying sediment, especially when the water table is high. Alterations to water courses, flooding, and fire can also follow in the wake of an earthquake. The most destructive earthquakes in recorded history are listed in Table 16-2.

According to the USGS, the role of earthquake prediction is "to give warning of potentially damaging earthquakes early enough to allow appropriate response to the disaster, enabling people to minimize loss of life and property." In this context, the aim is to specify a high probability that a specific earthquake will occur on a specific fault in a specific year.

Scientists, engineers, and architects know a great deal about earthquakes. The location of an earthquake can now be accurately determined, its magnitude can be measured, and structures can be built to withstand ground shocks with a minimum of damage. But predicting earthquakes still remains a major problem.

Three fundamental concepts underlie earthquake prediction:

TABLE 16-2
The Ten Most Destructive Earthquakes in Recorded History

Earthquake	Location	Magnitude	Date	Fatalities*	Principal cause of death
Shansi[1]	China	8	1556	Up to 800,000	Fires, landslides, floods
Sumatra	Indonesia	9	2004	280,000	Earthquake damage, tsunami
Tangshan	China	7.5	1976	255,000	Earthquake damage
Aleppo	Syria	Not known	1138	230,000	Earthquake damage
Damghan	Iran	Not known	856	200,000	Earthquake damage
Haiyuan (Gansu)	China	7.8	1920	200,000	Earthquake damage, landslides
Ardabil	Iran	Not known	893	150,000	Not known
Kanto	Japan	7.9	1923	140,000	Earthquake damage, fire, tsunami
Ashgabat	Turkmenistan	7.3	1948	110,000	Earthquake damage
Chihli	China	Not known	1290	100,000	Not known

* Approximate death tolls.
1. Shansi is the greatest natural disaster in all of recorded history.

▸ Earthquakes tend to be concentrated in broad geographic zones. This knowledge allows the identification of regions where earthquakes are most likely to occur.

▸ Earthquakes tend to recur along the same fault line in separate events. For example, when the strain released in one event is restored, then the process starts all over again, developing a kind of earthquake cycle. This cycle may not be exactly regular, but it does lead to the possibility that a number of earthquake events may, to some extent at least, be predictable.

▸ Earthquakes can occur in storms. Earthquake storms are an interesting phenomenon, and the theory behind them is that one earthquake can trigger a series of other earthquakes as the strain release moves along a fault line. Storms associated with the same event happen on a timescale of decades.

The theory behind earthquake storms is relatively recent and much of the data supporting the theory came from work carried out on two major transform faults found on land: the San Andreas Fault in the United States, and the North Anatolian Fault in Turkey.

Since 1936, the North Anatolian Fault, a 900-kilometer crack in the crust, has been the site of fourteen earthquakes with a magnitude exceeding 6.0. These earthquakes have occurred along a line moving progressively from east to west, each earthquake releasing stress at one part of the fault and passing it along the line to the next event. The earthquakes appear to come in storms, with a gap between them, as the fault ruptures progressively westward—unfortunately, from sparsely to densely populated regions.

Earthquake storms can have a major effect on individual civilizations as they strike an area like collapsing dominoes. The *earthquake storm theory* is somewhat similar to the generation of aftershocks in the same event, except now the separation between one earthquake shock and the next can be as long as decades. By contrast, aftershocks tend to follow a main shock on a much shorter timescale. One of the most important implications of the domino earthquake storm theory is that earthquakes are *not* random events in time, but are cyclic phenomena with one event triggering another.

Earthquake prediction covers a wide range of time spans: long-term (tens to hundreds of years), medium-term (years to months), and short-term (days to hours). A number of parameters have been used as potential indicators of earthquake activity:

▸ *Geological History*. Historical data provides a long-term footprint that can be a useful indication of the chances of an earthquake happening in a particular region. The historical view is helped considerably by the fact that earthquakes tend to occur in specific bands, although intraplate earthquakes are an exception. It must be pointed out, however, that although it is possible to identify *regions* that are susceptible to earthquake activity, such as the Pacific Ring of Fire, it is not generally possible to predict when earthquakes will occur there. Furthermore, we must never lose sight of the fact that low-probability earthquakes do happen, and can have disastrous consequences.

▸ *Seismic Data—Shock Waves*. To warn of earthquake activity, there are arrays of instruments designed to register seismic motion, and associated parameters, in a specific region. The principal instrument used in this monitoring is some form of a seismograph, an instrument that converts ground movement into an electrical signal that is recorded and processed with the aid of advanced computer systems. The zigzag signal recorded on a seismograph reflects changes in the intensity of ground vibrations. Scientists can derive a considerable amount of earthquake data from seismograms, such as the location of the epicenter, the focal depth, the amount of energy released, and sometimes the type of fault triggering the earthquake. Seismographs may be deployed at the surface or at depth in boreholes, and to achieve maximum payoff, they are often set up in network arrays around potential earthquake-generating faults, and the data acquired is then telemetered to a central processing HQ.

The sequence of seismic activity associated with the evolution of an earthquake cluster involves *foreshocks*, followed by the *main shock*, then the *aftershocks*. Both foreshocks and aftershocks can be useful in determining how to react to an earthquake event: Foreshocks are a warning to allow prequake preparations to be taken, and aftershocks are an indicator of how safe an area is for rescue services to enter and how long to keep them there. Foreshocks are a potential indicator of earthquake activity. However, a major problem with foreshocks is how to distinguish them from a genuine independent earthquake—in other words, how to tell if they are, in fact, foreshocks at all. The classic example is the earthquake that struck Haicheng, China, in 1975, a landmark event that might be considered to be the birth of modern scientific earthquake prediction. For a period of several months before the Haicheng event, there were indications of unusual activity in the area, such as changes in groundwater levels and odd animal behavior. Then increased seismic activity in the form of foreshocks was picked up and scientists were able to send out warnings days ahead of the earthquake that occurred on February 4, 1975, with a magnitude of 7.3. One response to the warnings was that, despite the cold weather, people were evacuated and remained outside their homes—actions that prevented many casualties when the earthquake struck and destroyed nearly 90 percent of the structures in the city. The final death toll in the Haicheng earthquake was around 2,000—a tragic loss of life, but only a small proportion of the estimated 150,000 casualties that would have occurred if no warning had been given. Unfortunately, the Haicheng event proved to be something of a false dawn for earthquake prediction, and in the following year the earthquake at Tangshan, also in China (see Panel 13), exhibited none of the foreshock precursors seen at Haicheng and resulted in the deaths of 250,000 people.

PANEL 13
THE TANGSHAN EARTHQUAKE*
(TANGSHAN, CHINA, 1976)

"One of the deadliest of all earthquakes"

Earthquake classification—"Ring of Fire," convergent oceanic/ continental plate margin.

Magnitude—sometimes reported as 8.1, but the Chinese govern-ment officially lists it as 7.8.

Death toll—255,000, but may have been as high as 500,000.

Principal cause of death—earthquake damage, compounded by subsequent tsunami.

Property loss—much of the city was destroyed.

▲

Tangshan is an industrial city and major population center situated in a coal-mining area facing the Gulf of Chihli (Bo Hai Sea) in the northeast of the People's Republic of China. The city is on a block of crust bounded by large faults. The Pacific Plate margin is being subducted under the Eurasian Plate, and the region was the site of several earthquakes in the 1960s and 1970s.

The 1976 earthquake started at 3:42 a.m. on July 28 and lasted for around fifteen seconds; the epicenter was about 8 kilometers directly below Tangshan. It was all the more devastating because it struck suddenly, without any appar-ent foreshocks, in the middle of the night when many people were asleep in bed. In those few brief seconds of terror, most of Tangshan over an area of 32 square kilometers was flat-tened, and more than 90 percent of the dwellings demol-ished.

This was earthquake damage on a grand scale. Those who had escaped being crushed or buried under rubble started a frantic scramble to locate and rescue survivors. But with the infrastructure in ruins and all lights and power gone, the odds against them in the darkness were over-whelming. Rescue workers attempting to reach the stricken city were bogged down as traffic clogged the only road still open. Those in Tangshan who were able worked heroically to set up emergency medical and food centers and saved a large proportion of those trapped under the rubble. Then, as if nature had not done enough, an aftershock with a magni-tude of 7.1 struck, killing many others trapped under the rubble by the major shock. Even then, it wasn't over, and rescuers entering the rubble of what had, only a few hours before, been a thriving city had to deal with the disposure of many thousands of corpses before the risk of disease added another dimension to the tragedy in human lives.

The earthquake had significant political repercussions in China, and the authorities were blamed for not having sufficient precautions in place, for inadequacies in their Resist the Earthquake, Rescue Ourselves initiative, and for their refusal to accept international aid. Nonetheless, the city of Tangshan was rebuilt and now has over a million citizens.

*The classification of the volcanoes, earthquakes, and tsunamis in the panels is based on post-plate tectonic thinking, and the terms used are described in the text.

▶ *Electromagnetic Measurements.* The electrical and magnetic properties of rocks can undergo changes as the crust suffers deformation. Instrumentation is deployed to detect and measure changes in the magnetic field, the electromagnetic field, and electrical resistivity in rocks around a fault. Disturbances to the magnetic field strength were noted in magnetometer readings prior to the Loma Prieta earthquake in California in 1988. The most apparent trend was an increase in extremely low frequency field strength, which was sixty times above normal a few hours before the earthquake struck.

▶ *Fluid Pressure Changes.* Changes in groundwater levels evidenced, for example, by the water in wells have been reported ahead of a number of major earthquakes, including Haicheng. These changes can be important earthquake indicators because over short time periods (a few days) the height of the groundwater table fluctuates in response to crustal deformation.

▶ *Odd Animal Behavioral Patterns.* One exotic approach to earthquake monitoring involves the behavior of animals prior to an event. For example, before the Haicheng earthquake, some local animals, including rats, chickens, snakes, horses, and cows, were reported to have behaved in strange, often agitated, ways. Several explanations have been put forward to account for this behavior. One suggests that the animals are responding to earthquake foreshock waves too small for humans to detect. Another claims the animals are reacting to low-frequency electromagnetic waves that increase before an earthquake. Animal behavior has often been dismissed as folklore by mainstream scientists, but there is no doubt it has been observed prior to earthquakes in several countries, including China, India, and Japan.

▸ *Seismic Hazard Maps.* They play an important role in tackling earthquakes. For a specific region, these maps are drawn by combining individual parameters, such as earthquake history and geological conditions, and what is known as "attenuation relations" that describe how ground motion is related to the magnitude of an event and the distance from the epicenter. Hazard maps provide seismic hazards data on the basis of probabilities that specific levels of ground vibration will be exceeded on, say, a fifty-year interval basis. The maps are used to formulate earthquake shaking levels that have a certain probability of occurring, and are designed to be slotted into building codes to aid engineers and architects in the design of buildings, bridges, dams, and other structures. The USGS produces hazard maps for the entire United States and many other parts of the world. These maps help to predict where an earthquake might happen. In addition, *predictive intensity maps* can be constructed to aid emergency response services to plan relief operations by indicating which areas have suffered most from an earthquake that has occurred very recently.

It is apparent that there are a number of parameters that can be used to broadly predict earthquake activity and so help to plan earthquake "damage limitation." But none of these parameters is entirely satisfactory, and there is a growing feeling among some seismologists that the problem is not that earthquakes are unpredictable, but that we simply do not know enough about their trigger mechanisms—even in the post–plate tectonics era.

In 1977, the U.S. Congress set up the National Earthquake Hazards Reduction Program (NEHRP), with the USGS as one of the four agencies involved. The prime aims of NEHRP are to provide seismic monitoring throughout the United States; to measure earthquake shaking, especially near active fault zones; and to disseminate earthquake event information, including, when possible, broadcasting early warnings before a significant earthquake event.

A number of individual national and international seismic monitoring networks can plug into NEHRP. One of these is the Advanced National Seismic System, which is a national monitoring system of more than 7,000 earthquake sensors that captures seismic event data and is designed to provide real-time earthquake information to emergency services. Another is the National Strong-Motion Project (NSMP), which employs strong-motion seismograph (accelerograph) instruments to record damaging earthquakes in the United States, with the aim of improving public safety in projects such as earthquake-resistant building design.

Many countries, and not just those such as Japan that are extremely earthquake-prone, now have seismic networks, and efforts have been made to set up some form of international framework for seismic data sharing. On an international scale, an important advance was made in the 1960s with the establishment of the Worldwide Standardized Seismograph Network (WSSN), essentially to monitor compliance with the 1963 aboveground nuclear test ban. Now, the Global Seismic Network (GSN) is in place to monitor seismic activity worldwide, with more than 136 stations in 80 countries on all continents. Established in 1986, the GSN uses state-of-the-art sensors to gather data that is routinely processed by the National Earthquake Information Center (NEIC) in Colorado and used to pinpoint the location, and measure the magnitude, of seismic events over the globe. In the United States, the USGS runs the recently (2006) reestablished National Earthquake Prediction Evaluation Council, a committee of twelve scientists mandated to advise the director of the USGS on earthquake prediction, forecasting, and hazard assessment.

In addition to these national and international seismic networks, a number of targeted earthquake prediction experiments have been set up with the aim of producing a model for earthquake prediction in a specific region. In the United States, the Parkfield Earthquake Prediction Experiment, which was started in 1984, was designed to use the San Andreas Fault as an earthquake laboratory. The rationale behind the experiment was that moderate-size (magnitude 6) earthquakes had occurred at fairly regular frequency (an average of every twenty-two years) in the area of the fault since 1857, and that another event resembling the historic earthquakes ("earthquake repeatability") was due sometime before 1993. The experiment employed a dense network of monitoring equipment designed to produce an integrated database on changes leading up to, during, and after an earthquake event. The idea was that the data would then be used to develop theoretical models for short-term earthquake prediction.

In the 1980s, data obtained during the experiment led to the prediction that a moderate-size earthquake would occur in the area of Parkfield, California, sometime in the period 1985–1993, with a 95 percent confidence level over this nine-year window. However, it did not happen, and an earthquake of the type predicted did not arrive until 2004, more than ten years outside the window. It was not only late, but the 2004 earthquake also differed in character from the predicted model—and in any case the prediction was thought to be too vague by some scientists.

Starting in the late 1970s, a Japanese site was selected for intensive earthquake study under the direction of the Japan Meteorological Agency (JMA). The site was the Tokai district, situated to the west of Tokyo. The aim was the collection of data that would eventually lead to the ability to predict earthquakes on a short-term basis, and to this end the JMA operates an earthquake observation network that feeds into an Earthquake Phenomenon Observational System (EPOS). The hope is that with the various data, short-term earthquake prediction of a Tokai earthquake will be possible.

Some events turn out to be landmarks, and two of these could be said to have changed the mindset of scientists with regard to predicting earthquakes. One was the prediction of the earthquake that didn't happen at Parkfield. The other was an earthquake that happened without any prediction. The Kobe Earthquake of 1995 had a magnitude of 7.2 on the Richter scale and was thought to have struck without any warning—although only later did scientists realize that precursors had come before the event. The earthquake in Kobe, Japan, resulted in the deaths of 5,200 people and caused very considerable structural damage. In fact, it was termed "the costliest natural disaster to strike any single country."

Both incidents—the event that didn't happen and the event that was thought to have happened without warning—shifted much of the emphasis away from earthquake prediction toward mitigation in the form of building construction and emergency response. In addition, a climate emerged in which the concept that earthquake prediction would ever be an attainable aim was seriously questioned. Underlying this somewhat pessimistic conclusion was the inescapable fact that viable global-scale earthquake precursors had simply not yet been identified.

So, where do we stand now?

Some level of earthquake prediction is undoubtedly possible. For example, using the most current data and the latest theories, the USGS has predicted that there is a 62 percent probability of an earthquake with a magnitude of 6.7 or greater occurring somewhere in the San Francisco Bay region before 2032. True, this warning is vague with respect to exact location and timing of the event, but it can be used for the implementation of damage reduction strategies.

Mitigation against earthquakes involves minimizing the effects of a disaster event. There is no doubt that in a vigilant society with sufficient financial resources, a great deal can be done to prepare for, and mitigate against, earthquake damage by adopting a long-term planning strategy that includes introducing building codes that incorporate specific safety measures. For example, earthquakes can result in up-and-down and side-to-side movements, both of which can damage structures. In areas where such earthquake damage is anticipated, however, suitable precautions can be taken with respect to *how* to build and *where* to build, and they can be enshrined in local regulations. Precautions of this kind can help to reduce both the death toll and the structural damage arising from an earthquake event, and they are often put in place with reference to earthquake "hazard maps." In this way, planners can insist by law on regulations that, as far as possible, ensure that structures are earthquake-resistant.

Civil engineers and architects have devised techniques that can be incorporated into building codes in earthquake-prone areas. These techniques include relatively simple modifications, such as using steel frameworks instead of wood, cross bracing with steel beams, and incorporating reinforced concrete into internal support walls in buildings. Such precautions provide resistance to shaking during earthquake shocks. Gas, electricity, and water pipes can be modified by using flexible jointing to allow for some movement without fracture. Other, more sophisticated precautions actually *allow* structures to move, or flex, during an earthquake event without causing damage. These precautions include inserting "base insulators" made, for example, of alternating layers of steel and an elastic substance like rubber, between the foundation of a building and the building itself; the effect of these insulators is to absorb some of the earthquake-derived motion that would otherwise cause damage. Some buildings require extra special precautions if they are to be constructed in earthquake-prone areas. Skyscrapers, for example, need deeper and more secure foundations and a stronger frame than would be required under normal conditions.

Disaster plans of varying degrees of sophistication are in place in earthquake-prone regions in many countries, including the United States, China, Japan, India, Pakistan, Greece, Turkey, and Russia. These plans involve national as well as local government, and in the developing countries, they are often drawn up in collaboration with international aid/relief agencies. Again, it must be stressed that it is important that the public be made aware of these plans, and that the plans themselves should be regularly tested.

An interesting earthquake disaster plan that involves a multistage "community collaborative" approach to earthquake recovery has been produced for the Tokyo area of Japan. The plan is published in the Tokyo Metropolitan Government "Recovery Manual" and includes the establishment of an "earthquake Restoration and Coordination Bureau" with the governor of Tokyo as its head, and the development of strategies for medical care, relief work, and building reconstruction following an earthquake event.

An example of a "relief /aid agency" approach to earthquake disaster management in collaboration with local partners was that used by ActionAid in the Pakistan earthquake of 2005 (see Panel 14). Working with local partners, other relief agencies, and the United Nations, ActionAid provided relief through a series of teams tasked with providing logistical coordination, giving advice and support to local partners, directing relief to the areas where it is most needed, and providing information and communication coordination.

PANEL 14
THE GREAT PAKISTANI EARTHQUAKE*
(KASHMIR, SOUTH ASIA, 2005)

Earthquake classification—convergent continental/continental plate margin.

Magnitude—7.6.

Death toll—75,000 to 80,000, with another 100,000 people injured.

Principal cause of death—burial under rubble.

Property loss—hundreds of thousands of buildings.

▲

The epicenter of the earthquake, which is also called the Kashmir or the South Asia earthquake, was located about 26 kilometers underground. The earthquake struck on October 8, 2005, and was followed by many aftershocks. The area lies in a collision zone between the Eurasian and Indo-Australian plates that gave rise to the Himalayas.

Damage from the earthquake was extensive, particularly in Pakistan-administered Kashmir and the North West Frontier Province of Pakistan, where whole towns and villages were destroyed. The shaking ground brought down buildings, and landslides ravaged villages and cut off roads.

Around 75 percent of the deaths occurred in Pakistan and Kashmir; in Muzaffarabad alone, close to the epicenter, 30,000 people are thought to have perished. The hazards arising from the earthquake were made more serious because of the remoteness of the region, the mountainous terrain, and damage to roads such as the Karakoram Highway. As a result of these various factors, it was difficult to bring relief to the 2.5 million people estimated to be left homeless.

It started to snow on October 13, and in an attempt to avoid further large-scale fatalities from exposure and cold and from disease, an appeal went out for survivors to congregate in the valleys and cities where help could be provided.

The Great Pakistan Earthquake provided an example of how local and international agencies respond to a disaster in a remote region. Although initially there was some delay in getting relief to the stricken areas, several local and international aid agencies succeeded, including the United Nations Development Programme (UNDP) and the United Nations Volunteers (UNV). At the same time, the Pakistan government organized its own national effort for the mass mobilization of volunteers. Organizations such as Oxfam carried out "quick response" relief by supplying blankets, tents, and water and basic hygiene kits to 200 refugee camps housing more than 500,000 people. Oxfam also provided help to rehouse victims on a more long-term basis. More than 100 other international agencies, such as Humanity First, the Red Crescent, the Red Cross, Christian Aid, Medecins Sans Frontieres (Doctors Without Borders), Muslim Hands, UNICEF, World Health Organization, and Save the Children, provided a variety of medical and shelter aid.

The earthquake also highlighted some of the difficulties involved in relief efforts on this scale. Problems that arise in these circumstances are often dominated by coordination difficulties between government and aid organizations. Early cooperation is needed, for example, to overcome problems in the transport and distribution of relief goods. This relief can be split into two broad categories: 1) short-term and immediate on-the-ground relief, often carried out by aid agencies and charities in tandem with local providers, and 2) long-term strategic planning for the future, often involving the major aid agencies and tending to concentrate on bringing in resources for building a viable infrastructure.

A great deal was achieved in the immediate aftermath of the Great Pakistani Earthquake, but such was the scale of the disaster, augmented by difficult remote terrain, that one year later 1.8 million people were still housed in temporary shelters. Lessons should be learned from the experience; some of them quite basic, such as the need to supply tents that are suitable for use in winter conditions.

Interestingly, in addition to bad terrain, one reason for the slow response cited by several of the aid agencies was the poor media coverage at the start of the emergency—a criticism that could not be leveled at the 2004 Indian Ocean tsunami (see Panel 15).

*The classification of the volcanoes, earthquakes, and tsunamis in the panels is based on post-plate tectonic thinking, and the terms used are described in the text.

The USGS has produced a series of earthquake hazard maps that show the distribution of earthquake activity that has a certain probability of occurring in the United States. One of these maps for the U.S. mainland is reproduced in Figure 16-1.

Three very broad trends in earthquake distribution can be identified from the map, although it must be remembered that there are a number of caveats with this approach, since some of the data used in constructing the map is taken from historical records.

▸ The area with the highest hazard risk is along the West Coast of the continent, including some regions of Alaska. These areas are part of the Pacific Ring of Fire, and Alaska and California are 1 and 2, respectively, in league table of U.S. states that suffered earthquakes with a magnitude greater than 3.5 in the period 1974–2003. In fact, it was in Prince William Sound, Alaska, that the strongest earthquake to strike the United States in recorded history (magnitude 9.2) occurred in 1964. This earthquake zone also includes the seismically active San Andreas Fault.

▸ Areas in the central United States have the lowest seismic activity.

▸ The eastern United States has a number of earthquake activ-

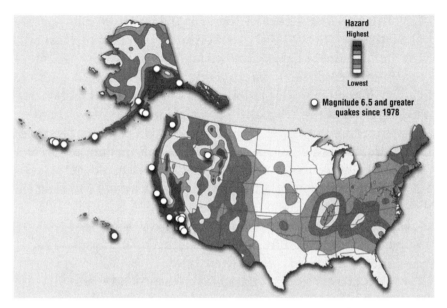

Figure 16-1. The distribution of earthquake activity in the United States, as shown on a seismic hazard map. The area with the highest hazard risk is along the West Coast of the continent, including some regions of Alaska. Areas in the central United States have the lowest seismic activity. The eastern part of the country has a number of earthquake activity concentrations; in particular, the oval-shaped New Madrid Seismic Zone (see Panel 6). Credit: United States Geological Survey.

ity concentrations; in particular, the oval-shaped New Madrid Seismic Zone (see Panel 6). Another is located in the northeast United States.

In the United Kingdom, the prediction of earthquake hazards is part of the remit of the British Geological Survey (BGS). In the eyes of the public, the UK is not a particularly earthquake-susceptible region in the sense that, for example, California or Japan are. But although seismic risk is low, the level of seismicity when it does occur is moderately high and is sufficient to pose a potential hazard to sensitive installations such as chemical plants, dams, and nuclear power stations.

Historically, the study of earthquakes in the UK was essentially in the hands of amateur seismologists such as the Victorian E. J. Lowe and later Charles Davidson and A.T.J. Dollar. The BGS became strongly involved in the 1970s with the establishment of

LOWNET, a seismic network situated in the Lowland Valley of central Scotland, run by the Global Seismology Group of what was then the Institute of Geological Sciences. The network was subsequently expanded into a countrywide seismic monitoring system that runs 146 monitoring stations across the United Kingdom. The network can detect earthquakes with a magnitude greater than 1.5, and typically records between 300 and 400 events per year, of which approximately 10 percent will have a magnitude greater than 2.0. Later, concerns for the safety of nuclear power stations led to the production of a comprehensive earthquake catalog for the UK drawn up by the BGS and other bodies.

The geographical distribution of earthquakes in the UK shows a number of very broad features (see Figure 16-2):

▸ In Scotland, most earthquakes are concentrated on the west coast.

▸ In England and Wales, most earthquakes are concentrated in two broad bands essentially parallel to the west coast, with the highest activity being in the southeast of the country and lowest in the northeast, which is a "quiet zone." Earthquake activity is also found in the English Channel and off the Humber estuary. Within the overall picture, there is a hint of a pattern because a number of centers appear to show persistent earthquake recurrence. They include northwest Wales, which is one of the most seismically active areas in the UK. It was here that one of the largest ever UK earthquakes with an epicenter on land occurred in 1984 with a magnitude of 5.4. Two other large earthquakes with magnitudes estimated to be about 5.75 occurred in 1382 and 1580, with epicenters in the area of the Dover Straits. The biggest earthquake ever to strike the UK occurred in 1931 and had a magnitude of 6.0. However, perhaps the most striking feature of the earthquake distribution in the UK is that it appears to bear no relation to the structural geology of the landmass.

▸ Ireland appears to be almost free of earthquakes.

Using a statistical approach, the BGS reports that, on average, the United Kingdom can expect an earthquake of magnitude 3.7, or larger, every year; an earthquake of magnitude 4.7, or larger,

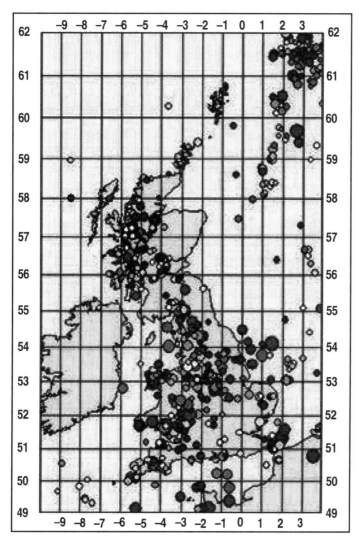

Figure 16-2. The distribution of earthquakes in the United Kingdom. A number of very general trends are apparent. 1) In northernmost Scotland, most earthquakes are concentrated on the west coast. 2) In England and Wales, most earthquakes are concentrated in two broad bands essentially parallel to the west coast. 3) Ireland appears to be almost free of earthquakes. Credit: British Geological Survey.

every ten years; and an earthquake of magnitude 5.6, or larger, every 100 years.

Public education plays a lead role in mitigating against the effects of earthquakes, and the USGS and the BGS are both active in this field. The Earthquake Hazards Program run by the USGS distributes information on a wide range of earthquake-related topics, including the basic science behind earthquakes. The BGS provides a variety of seismic hazard services targeted at both local and international levels. In particular, the BGS's *A Short Guide to Seismic Hazards* is a useful public education tool.

As well as preparing people via education programs, public communication is vital during and after any disaster event in order to mitigate suffering and prevent the spread of disease. For example, in response to the 2005 Pakistan earthquake, an Earthquake Relief and Rehabilitation Communication Plan was set up to create public awareness and to encourage information sharing between different government and outside agencies, especially in the areas of health, water and sanitation, and child protection.

Despite all such efforts, however, it must be remembered that assessing earthquake hazards can be a very misleading exercise. A scientifically generated prediction may genuinely conclude that the probability of a damaging earthquake striking a specific area is extremely low. Unfortunately, this "extremely low" possibility may become a nightmare. Perhaps it is apt to conclude by quoting the words that have been attributed to Charles F. Richter, the man who devised the Richter scale: *No one but fools and charlatans try to predict earthquakes.*

It is possible that earthquake "damage limitation" may be the most we can expect for the near future.

TSUNAMIS: "THE BIG WAVE"

The three key factors that determine the destructive power of a tsunami are the force of the event that triggered it (e.g., the magnitude of an earthquake), the distance the tsunami wave has traveled from the trigger point, and the topography of the shoreline it hits. As a result, tsunamis are usually most destructive closest to the site of generation, and since 1850, they have been responsible for the deaths of more than 400,000 people.

When tsunamis strike the physical barrier of a shoreline their height increases dramatically and their speed carries vast volumes of water forward onto the land. The coastal areas hit by a tsunami can span the range between built-up population centers and pristine, poorly populated environments. In pristine coastal regions, the major impact will be on species habitats. But inevitably it is in developed coastal areas that tsunamis are most catastrophic, and the damage here may include extreme loss of life, destruction of buildings, and the breakdown of the infrastructure of society. It must also be remembered that because it can traverse entire oceans, the same tsunami can strike several countries and a wide variety of habitats.

All low-lying areas are at risk, and tsunamis can cause a spectrum of hazards, the most devastating usually being impact damage, inundation, drag-back, and erosion.

Impact damage arises directly from tsunami waves striking a coastline. The waves often come in a series and can reach heights of up to 50 meters. They travel faster than a person can run and break with a devastating force. The impact is compounded by acquired debris (e.g., large rocks, cars, and even ships) that is swept along and adds to the destructive power as tsunamis crash into structures. The impact damage can produce loss of life and have a number of other effects—such as damaging ships in harbor, destroying buildings, and bringing down power and telephone lines.

Inundation is the flooding of low-lying regions, when the tsunami pushes large quantities of water above the normal high-water mark. This is termed *run-up,* and it can create as much damage as the impact of a large tsunami wave. Sometimes, tsunamis move upstream in rivers, extending the area of damage.

Even if people and animals escape the initial effects of a tsunami wave as it strikes a coast, they can still be affected by *drag-back* as the wave withdraws and water is dragged back out to sea.

Tsunami waves can cause *erosion* to the foundations of structures such as buildings, bridges, railway lines, and dams. Erosion will either destroy the structures immediately or cause long-term damage that renders them unsafe and requires rebuilding. Tsunamis can also cause fires—for example, from collapsed power lines, or from damage to facilities such as oil storage tanks, chemical plants, and nuclear power stations.

Tsunamis can damage the natural environment in various ways. Offshore, they can cause considerable harm to coral reefs and other shelf ecosystems. Onshore, they can damage mangrove and other coastal forests. They can also destroy wetland habitats and farm-land by flooding with seawater and by contamination with any pollutants they may have picked up.

Less apparent, but equally important, is that both tsunamis and earthquakes can damage all parts of the fabric of society, and whole regions of population can be left rudderless. The psycholog-ical effects on a region that is left structureless and totally disorien-tated can be profound. In fact, tsunamis have been cited as a strong contributing factor to the destruction of the Minoan civilization that was once centered in the Mediterranean.

Although the origin of a tsunami is related to a specific trigger, it must be remembered that the tsunami wave that is generated can spread over vast ocean distances. In this way, damage can be caused to sites thousands of miles away from the region in which the tsunami was born.

Tsunamis can strike swiftly, without warning, in widely scat-tered places, so that predicting them presents a great challenge. Many tsunamis are almost too small to be picked up, except by very sensitive instruments, but harmful tsunamis probably occur at the rate of about one or two every year. In the last decade, de-structive tsunamis have been reported in South America, Central America, Indonesia, the Philippines, Papua New Guinea, Turkey, and Japan. The most destructive tsunamis of all time are listed in Table 16-3.

Two things must be remembered when we think about the ori-gin of tsunamis. First, most tsunamis are caused by undersea earth-quakes, but not all earthquakes cause a tsunami—there appears to be a cutoff magnitude, thought to be around 6.5 or 7.0 on the Rich-ter scale, that has to be exceeded before a tsunami can be triggered. Second, around 90 percent of all tsunamis originate in the Pacific Ocean, mostly in the Ring of Fire, an area rich in tectonic activity.

Because most tsunamis are caused by earthquakes, and because there is no satisfactory way of predicting earthquakes themselves, it follows that there is no reliable way of forecasting a tsunami or working out the probability that a tsunami of a particular size will strike a specific area at a specific time. As a result, the emphasis

TABLE 16-3
The Ten Most Destructive Tsunamis in Recorded History

Tsunami	Location	Trigger	Date	Fatalities*
Indian Ocean	Indonesia	Earthquake	2004	225,000 or more
Mediterranean Sea	Messina, Italy	Earthquake	1908	Possibly more than 100,000 from earthquake and tsunami
Mediterranean Sea	Several Greek islands	Earthquake	~1400 BC	As many as 100,000
North Pacific	Genroku, Japan	Earthquake	1703	As many as 100,000
North Atlantic	Lisbon, Portugal	Earthquake	1755	60,000
North Pacific	Taiwan	Earthquake	1782	40,000
Indonesia	Krakatau	Volcano	1883	36,000
North Pacific	Tokaido, Japan	Earthquake	1707	30,000
North Pacific	Sanriku, Japan	Earthquake	1896	26,000
South Pacific	Northern Chile	Earthquake	1868	25,000

* Approximate death tolls. Tsunamis are triggered by earthquakes and volcanoes, and it is often difficult from the records available to allocate fatalities to one specific cause. In compiling this table, tsunamis have been selected if they have made a significant contribution to the overall death toll of a disaster event.

must shift and be focused on the design of warning systems that monitor the *progress* of a tsunami after it has been generated, and on setting up programs that *mitigate* against the effects of a tsunami.

The time taken for a tsunami to strike a coastline depends on the distance of the landmass from the point of origin of the trigger event. Coastlines close to the origin of a tsunami can be devastated in hours, or even minutes, of the initial disturbance, so that no warning is going to be possible anyway. However, because tsunamis transmit their destructive power on an oceanwide scale, albeit at a great speed, warnings can, at least in theory, be given to countries that are likely to be hit as the tsunami travels. Ideally, these countries should have in place some form of public alarm system to warn the population of the approaching tsunami, because without a mechanism for dissemination even the most accurate information will be useless. But *if* advanced warning systems and disaster plans are in place, then tsunami death tolls can be reduced, sometimes to a major extent, by even the simplest of precautions— such as moving populations to higher ground or evacuating ships from harbors to deep water.

The actual speed at which tsunamis travel varies from one event to another, but as a rough guide it takes something between ten and twenty hours for a tsunami to cross the Pacific Ocean. The Indian Ocean is smaller than the Pacific, and cross-basin travel times are shorter, probably no more than twelve hours. In the 2004 Indian Ocean tsunami (see Panel 15), which had a trigger point off the west coast of Sumatra, the wave reached the coast of Indonesia in thirty minutes, Thailand in eighty to 100 minutes, Sri Lanka and the east coast of India in 120 minutes, Somalia in seven hours, Kenya in ten hours, and the tip of South Africa in twelve hours (see Figure 16-3). These travel times should have allowed at least some degree of warning to be given ahead of the tsunami striking coastlines in Thailand and beyond. But no warning system existed. One reason is that the perceived risk of a large-scale Indian Ocean tsunami was not high. But there is little doubt that a warning system would have saved many thousands of lives in countries like Sri Lanka, where the death toll was 30,000, and the east coast of India, where 10,000 died.

Hawaii has had a warning system since the 1920s, and more sophisticated systems were developed in some areas of the Pacific in response to tsunami events in 1946 and later in 1960. The Tsunami Warning System in the Pacific program was established in 1966 and had inputs from the Pacific Tsunami Warning Center and the International Tsunami Information Center, both in Honolulu.

Any tsunami monitoring program should be based around four elements:

▸ Recognition that an earthquake has occurred at a specific location, usually through the use of a seismograph network

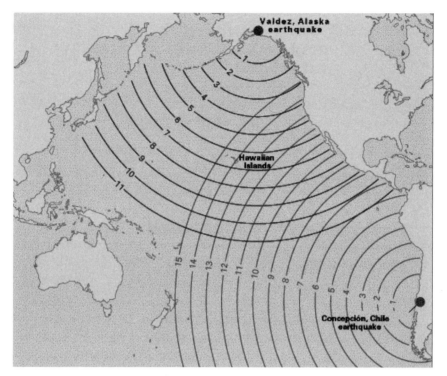

Figure 16-3. Tsunami travel times in the Pacific Ocean. Travel times (in hours) are shown for the tsunamis produced by the 1960 Concepción, Chile, earthquake and the 1964 Good Friday, Valdez (Anchorage), Alaska earthquake. Credit: United States Geological Survey.

▸ Distribution of warnings to all countries that will potentially be in danger if a tsunami follows the earthquake

▸ Sending out either an alert, if a tsunami has been generated, or a stand-down, if one has not

▸ Tracking a confirmed tsunami as it crosses an ocean and, equally important, constantly updating information on its progress to the countries in danger

Recently, significant advances have been made on two important fronts: the development of real-time, deep-ocean tsunami detectors and the establishment of protocols for tsunami information sharing on both regional and international scales.

PANEL 15
THE INDIAN OCEAN, OR ASIAN,
EARTHQUAKE AND TSUNAMI*
(INDONESIA, 2004)

"The mother of all tsunamis"

Tsunami classification—earthquake generated.

Location of earthquake—off the west coast of Sumatra.

Magnitude—9.0.

Tectonic setting—Although sometimes referred to as being in the Ring of Fire, because the extreme western edge of the ring extends into this region of the Indian Ocean, the Sunda Strait is an area of subduction in the Alpide earthquake belt.

Death toll—225,000, but could be as high as 300,000.

Principal cause of death—the tsunami.

In many ways the Asian tsunami of 2004 was a benchmark that caught the global imagination, as TV images of almost-unimaginable death and devastation were transmitted around the world in what was almost real time. It was also a wake-up call, alerting a stunned public to the very real dangers of a large tsunami.

It was not that people thought they were safe in an age of scientific triumph; it was the sheer scale of the loss of life and the destruction of property that was overpowering. This was a natural force striking with a power that was impossible to comprehend. Never had the expression *nature in the raw* been more apt. Yet the tsunami brought out the best in those that viewed it from afar, and record sums of money were raised for the aid and relief of the hundreds of thousands who had fallen victim to the catastrophe.

The earthquake struck at seven in the morning local time on December 24, 2004, and shocks were felt across the entire planet. Twenty minutes later the tsunami struck Banda Aceh, Indonesia, and within hours the shores of Thailand, Sri Lanka, and India were battered by tsunami waves as high as 15 meters.

In the area where the earthquake was generated the Indian, Australian, and Burmese plates collide, and the trigger occurred at the interface between the Indian and Burma plates off the west coast of northern Sumatra. In the Sunda Trench, which marks the interface between the two plates, a 1,000-kilometer section of the plate boundary fault line slipped. It happened in two phases spread over several minutes. The first phase in the south involved a 400-kilometer-long, 100-kilometer-wide rupture, moving at a speed of several thousand kilometers an hour in a northwest direction 30 kilometers below the seabed—the longest rupture ever known to have been caused by an earthquake. The second phase, in the north, proceeded more slowly, and here the fault changes from a subduction to a strike-slip type. The faulting associated with the earthquake lasted for up to ten minutes and became the longest on record. It made the entire earth vibrate and triggered other earthquakes. Some of the smaller islands in the region actually shifted position by a few inches.

The overall result of the massive ruptures caused the seabed to rise by a few meters and change its shape. The movement displaced approximately 30 cubic kilometers of seawater and triggered a series of tsunamis that struck almost all the coastal locations in the Indian Ocean, including Indonesia, Sri Lanka, Thailand, India, Myanmar, the Maldives, Malaysia, the Seychelles, Somalia, and Kenya. Waves of up to 25–30 meters tall were recorded in some areas.

The tsunami generated by the earthquake crossed the ocean basin and struck the countries around the Indian

Ocean in a domino effect, as it spread out from the region of the epicenter and changed from a local to a distant tsunami. In some regions there was a lag time of several hours between the earthquake and the tsunami. Almost every country around the Indian Ocean was taken completely by surprise. The principal reason was the absence of a tsunami warning system in the Indian Ocean, and the fact that inadequate plans to inform the public of the impending disaster were in place. Since the events of 2004, warning systems have been set up in the Indian Ocean.

One of the most catastrophic features of the Asian tsunami was that it struck in a series of waves, so that as soon as people thought they may have escaped the worst, another wave came crashing down in a cycle of "retreat and rise," in which it was the third wave, not the first, which was the strongest.

With a death toll of up to 300,000, one-third of them children, and more than one million people displaced from their homes, the Asian tsunami of 2004 is one of the greatest natural disasters in the history of the world. Everything about this earthquake-tsunami system was on a massive scale, and according to the United Nations, the relief operation was the most costly ever mounted. In the immediate aftermath of the tsunami, aid was provided for medical services, food, and shelter, with the first requirement being for clean water and sanitation to avoid diseases such as typhoid, diphtheria, dysentery, and cholera taking hold. Once the immediate relief aims had been met, psychological counseling was provided for people who had lost relatives or been displaced from their homes. Full reconstruction could take up to ten years.

But the tsunami caught the imagination of the world and set off one the greatest relief responses of all time, with private individuals, acting through the aid charities, and governments together raising funds totaling billions of U.S. dollars.

*The classification of the volcanoes, earthquakes, and tsunamis in the panels is based on post–plate tectonic thinking, and the terms used are described in the text.

An earthquake can be detected on a seismograph a few minutes after it has happened, but confirmation that a tsunami has actually been generated must come from instruments that measure water movements and waves. These instruments are deployed in both coastal surface locations (tidal gauges) and open-ocean, deep-water locations (buoys).

A tidal gauge is an instrument for measuring sea level relative to a nearby geodetic standard. In the early versions, tidal gauges consisted essentially of a tube, or stilling well, that calmed the water around the measuring device. The tube, which was typically about twelve inches wide, contained a cylindrical float that hung from a wire. Seawater entered the gauge and the float rose up and down with the tide. Today, however, the modern tidal gauge contains electronic sensors and measurements are recorded on a computer, and an important advantage of a modern tidal gauge system is that sea-level data can be transmitted to a central collection facility to provide *quasi*-real-time data. Although primarily designed to provide data on tides, tidal gauges can be incorporated into tsunami warning systems.

By placing sensors on the bottom of the sea, it is possible to identify large tsunami waves that extend down the water column. A major breakthrough in the design of instruments for tsunami detection came with the development of the Deep-ocean Assessment and Reporting of Tsunamis (DART) buoy system. The DART buoy arrays are designed for the early detection and real-time reporting of tsunamis in the open sea away from the landmasses by tracking and measuring a tsunami wave as it travels across the ocean. The DART system consists of a bottom pressure recorder (a *tsunameter*), anchored to the seafloor, which is deployed together with a sea surface buoy, for the real-time reporting of data. The bottom recorder and the surface buoy remain in communication via an acoustic link. The bottom pressure recorder is capable of picking up a tsunami as small as one centimeter, and data retrieved from the seabed is relayed through the surface buoy and then via a satellite to ground stations. The system normally operates in "standard" mode, in which data are reported as the average of four fifteen-minute values, but it switches to "event" mode, during which data is supplied over shorter periods, when a tsunami event is picked up. The DART project is part of the U.S. National Hazard Mitigation Program (NTHMP) and has proved to be an efficient tsunami warning system. In the DART network, buoys are positioned mainly around the Pacific Ocean, but they are also deployed in a number of other oceanic regions.

The need for tsunami warning coverage to be extended beyond the Pacific was highlighted by the Indian Ocean tsunami of December 2004. This wake-up call resulted in a decision to reevaluate the global tsunami warning system and extend it to cover regions such as the Atlantic Ocean, the Caribbean, the Mediterranean Sea, the Black Sea and the China Sea, as well as the Indian Ocean. This task is overseen by UNESCO's Global Strategy for the Establishment of Tsunami Early Warning System.

As a result of changes made over the past few years, we are now moving toward reaching the aim of the Intergovernmental Oceanographic Commission (IOC) to establish a global network of mixed *tidal gauge/deep-sea buoy* tsunami warning systems. On an international level, the National Oceanic and Atmospheric Administration (NOAA) has incorporated a network of tidal stations into the Pacific tsunami "detection and warning network." Together with DART data, this network will aid tsunami detection and tracking. In the Indian Ocean, efforts have been made to set up a tsunami detection/monitoring system by linking earthquake data and tidal gauge/buoy data, to provide tsunami impact times. The Indian Ocean Tsunami Warning and Mitigation System (IOTWS) was established under the auspices of UNESCO eighteen months after the 2004 tsunami. The system consists at present of a network of twenty-five new seismograph stations that are linked in real time to twenty-six national analysis centers and three DART buoys. In addition, plans are underway to extend tsunami warning systems to the Atlantic and the Mediterranean Sea.

From the point of view of potentially harmful effects, tsunamis can be thought of as being examples, albeit often extreme examples, of "tidal surge waves," and a number of precautions can be taken to mitigate against their impact.

Computer modeling is used to investigate various aspects of tsunamis. For example, the MOST (Method Of Splitting Tsunami) model simulates three stages in the life of a tsunami: *earthquake generation, oceanic propagation,* and *coastal inundation.* Models of this kind are useful in tracking tsunamis, especially distant events that have travel times of several hours. Equally important is the extent to which a tsunami will affect a local region, and central to any tsunami mitigation plan is the production of *inundation maps*. These are drawn up using a combination of factors such as historical events, distance from sites of potential earthquakes and volcanoes, seafloor bathymetry, and onshore topography. Such data is fed into models of various tsunami scenarios, which may include the worst-case example. Tsunami inundation maps identify the areas most likely to be struck by a tsunami event, and so these maps are vital to any long-term planning in tsunami-susceptible environments. The maps are also used in the planning of population evacuation routes.

Once areas at risk have been identified, precautions can be put in place. The precautions are aimed at saving lives and protecting property both onshore and offshore. There are three general types of tsunami precautions:

▸ *Off-Shore Protection.* Certain precautions are aimed at offshore protection from tsunamis by reducing the energy of the waves. These include constructing defenses such as breakwaters, seawalls, and flood barriers. One particular type of flood barrier or gate is designed to prevent a tsunami rushing upriver, as happened in the 1755 Lisbon tsunami (see Panel 5).

▸ *Public Safety and Evacuation.* To directly safeguard the public, one of the most important precautions is the evacuation of the population. This can involve either the complete evacuation of the area at risk using well-planned evacuation routes, or if total evacuation is not possible, a partial evacuation by moving people to safe sites. Safe sites include tsunami-resistant towers or shelters, and natural, or artificial, preidentified high ground.

▸ *Land Use Planning.* In tsunami inundation zones, local laws may ban some facilities, such as schools, hospitals, and certain industrial plants, from being built in these susceptible zones. For structures that have to be built in these zones, special building regulations can force constructors to take various measures to strengthen structures. The way the stronger buildings can survive tsunamis was highlighted by one of the most enduring worldwide media images of the 2004 Indian Ocean tsunami, which showed several mosques that remained standing after the tsunami had struck, alone among the wilderness of destruction.

Finally, another factor should be considered in relation to tsunami mitigation, and that is curtailment of the deliberate destruction of natural barriers to tsunami waves. Barriers of this kind include sand dunes, but of more importance are mangrove swamps and coral reefs. They form natural buffers between the open sea and the area immediately inland of the coast, and both can break the force of a tsunami before it can move onshore. And both are under considerable threat.

Mangroves are mainly located in coastal saline environments, particularly estuaries, in the tropics and subtropics, and once covered 75 percent of the coastline in these areas. With their mass of tangled roots, much of which is above water, mangroves are very effective at slowing down and breaking up a tsunami that is approaching a shoreline. Unfortunately, in some Southeast Asian countries, mangrove deforestation has been used to free up the land for other purposes, and a number of estimates suggest that as much as 80 percent of the mangroves have been destroyed in these areas. In locations where they still flourish, mangroves are reported to have saved, or at least played an important part in saving, many lives in the Indian Ocean tsunami of 2004 in Indonesia, India, Sri Lanka, Thailand, and Sumatra.

Coral reefs can mitigate against a tsunami by cutting down its force before it hits a shoreline. There was evidence that coral reefs offered some degree of protection against the 2004 Indian Ocean tsunami, although not as much as that provided by mangrove swamps and coastal forests. The coral reefs themselves come under environmental pressure from tsunamis and can suffer severe damage, although it is thought that many reefs that were attacked in 2004 will recover in less than ten years. Unfortunately, coral reefs are under global attack from a number of sources, such as climate change, ocean acidity, and pollution.

The U.S. agencies involved in predicting, monitoring, responding to, and fostering the awareness of tsunamis are NOAA and the USGS. Systems have been put in place to help minimize loss of life and property from tsunamis on local, state, and national levels. So, how might a local tsunami warning /mitigation system actually work? This question can be answered in a very general way by using the example of a hypothetical U.S. tsunami warning system and the process by which it might be switched on following an earthquake in the Pacific Ocean.

▸ Earthquakes are monitored by a network of seismic instruments (see Chapter 5) that provide vital information on where and when a tsunami might be generated. Initial tsunami "warning zones" can then be identified based on the seismic data, and related to the magnitude of the trigger earthquake. For example, for earthquakes with a magnitude greater than 7.0 on the Richter scale, the warning might cover the coastlines within two hours estimated tsunami travel time from the epicenter, and for an earthquake of 7.5 magnitude, this warning would extend to three hours travel time. Initial warnings of a potential tsunami can then be issued to coastal areas in danger.

▸ Changes to deep ocean waters are monitored mainly using DART information, which is transmitted to the tsunami warning centers. Based on the incoming data, a decision will be made on whether or not a tsunami has actually been generated by the earthquake. In the absence of a tsunami, the initial warning bulletins will be canceled. However, if the tsunami warning is confirmed, further bulletins will be issued and authorities, such as state and local officials responsible for danger management, will be alerted in the coastal areas that may be in danger. Emergency managers should be able to make use of tsunami inundation maps prepared for many regions in danger from tsunamis. The maps are compiled using the latest computer techniques. Besides indicating specific areas that are potentially at risk from flooding, the maps are also used to identify evacuation routes and aid long-term local planning.

▸ NOAA Weather Radio will broadcast Emergency Alert System tsunami warnings to the general public via radio and TV, to reach homes, schools, businesses, and health facilities and to activate plans for the evacuation of low-lying areas. "All-hazard alert broadcasts" will also be put out to remote coastal areas.

Essential to all the risk planning are tsunami education programs, such as those run by NOAA and the USGS in the United

States. Their overall aim is to build up a "tsunami resilient community." This concept envisages a community that is alert to the risks associated with tsunamis, aware of emergency plans for the evacuation of danger areas, and educated in how to respond to a tsunami event—for example, by securing private property. In addition, NOAA has a "tsunami ready" program that promotes tsunami awareness and readiness in terms of cooperation among federal, state, and local management agencies and the public.

The U.S. position with respect to tsunamis was presented in a report on "Tsunami Risk Reduction for the United States: A Framework for Action," prepared by the National Science and Technology Council in 2005. In effect, this report proposed a plan for the development of tsunami-resilient communities by improving federal, state, and local capabilities in a number of areas. These areas included determining the threat from tsunamis; setting up effective early warning systems; sharpening preparedness, mitigation, and public communication; and expanding research and international cooperation. In the same year, the United States announced a plan for an improved tsunami detection and warning system designed to provide the country with an almost 99 percent tsunami detection capability. It will work as part of the Global Earth Observation System of Systems (GEOSS)—a sixty-one country, ten-year initiative to revolutionize our understanding of how the earth works.

In the United Kingdom, an interagency report was commissioned, and in 2006 it was published under the title, "The Threat Posed by Tsunami to the UK." One of the cooperating agencies was the BGS, which provides natural hazard scientific data to the UK government and is involved in a public education program. The report concluded that tsunami events in the northeast Atlantic and the North Sea are possible, but rare, and that they could be contained by current defenses that have been designed to resist storm surges. It was noted, however, that tsunamis pose a different kind of threat and that, in particular, they would not have the meteorological advanced warning that storm surges have. In the light of this finding, a program of research was recommended to identify the impact "envelope" likely to be at most risk from earthquake tsunamis, to look more closely at the different impacts that tsunamis and storm surges would have on coastal regions, and to study typical impacts of near-coast events and identify their potential hazards. Following this research program, public awareness and emergency planning activities are to be reviewed. The report concluded that because the events are so infrequent, a tsunami early warning system could only be sustainable if it was also used for other purposes.

COPING WITH NATURAL DISASTERS

Where do we stand now in relation to monitoring, predicting, and coping with volcanoes, earthquakes, and tsunamis?

With the advent of plate tectonics theory, we understand the science behind the generation of volcanoes, earthquakes, and tsunamis, and with new and better technology we can monitor them to an unprecedented extent. The instrumentation developed, such as DART buoys, are being combined into global networks, and real-time data is now being shared on a scale not known before the wake-up call of the 2004 Indian Ocean tsunami. But we are better at *monitoring* than *prediction,* and there is still a long, long way to go before scientists can, if indeed they ever will be able to, accurately predict the occurrence of, say, a major earthquake. But one thing has become blatantly apparent, especially since the events of 2004—and that is the need not only to develop better monitoring networks, but also to bring the public on board.

Society must approach disaster planning in two ways. The first is the *big picture* preparedness of the authorities, which requires them to have monitoring systems and major disaster plans in place. These plans involve a range of emergency services and are regularly tested. The second, and no less important, element is the *small picture* preparedness of the public, which will allow homeowners, schools, businesses, and health facilities to cope with a disaster.

The degree of public awareness necessary to achieve "small picture" preparedness requires education at all levels, together with a mechanism for disseminating information. This information should include what to do at the personal and family levels when disaster strikes. For example, in California, a series of leaflets are available from the Governor's Office of Emergency Services to cover even the most basic precautions. One leaflet is entitled *Emergency Supplies Checklist,* and details the preparations that should be made in the home in order to survive the vital seventy-two hours after a disaster strikes—a time when power and water supplies and the phone system may be out of action and emergency services will be stretched to the limit. The motto in California is "Be Smart, Be Prepared, Be Responsible," and under this umbrella another guidance plan is available called *10 Ways You Can Be Disaster Prepared*. It covers such topics as identifying risks, creating a family disaster plan, and preparing children for what to do.

Public preparedness programs are also run in other disaster-prone areas outside of the United States. Shizuoka Prefecture in Japan issues an *Earthquake Disaster Prevention Guidebook* outlining an eleven-point plan covering all aspects of how to respond to earthquakes. All this public involvement is designed to establish a *disaster resilient* population that, among other benefits, will make the role of the authorities much easier as they attempt to cope with difficult disaster scenarios.

▲

There is no doubt that the theory of plate tectonics has led to a better understanding of the science that underpins volcanoes, earthquakes, and tsunamis. But has the theory provided a better way of predicting the disasters they bring? At present, we are still a very long way from accurate predictive protocols for any of the trinity of natural disasters. Furthermore, our knowledge of plate tectonics is very far from complete. We know the broad brush-strokes, but a great deal of detail is still hidden. As it is brought to light, perhaps it will provide data that can be used to refine the predictive protocols.

But whatever the eventual outcome, so far the road to plate tectonics has taken us on a journey that has stretched the boundaries of our knowledge and, in doing so, changed the way we view the earth.

EPILOGUE
Deadly Years and Megadisasters

CHANGE

As myth gave way to science and science fought to free itself from the shackles of religious dogma, the picture that finally began to emerge was one of a planet that was old and that had, from its birth, undergone extreme but slow change. Nothing was permanent; and that notion, perhaps more than any other, shocked the Victorians and rattled the foundations of their settled world.

It was recognized that geological strata, with their folding and uplift and their fossil communities, were leaves in a book that went back in time and recorded many of the changes that the planet had suffered. Eventually, at the end of a long road, the theory of plate tectonics was able to explain how these changes occurred as plates shifted against each other, causing strain to build up in the lithosphere and releasing molten rock from the depths of the earth as the planet continued to evolve. As a result, the changes came at a cost because they brought with them the specter of disasters—particularly those arising from volcanoes, earthquakes, and tsunamis.

THE YEAR 2004

According to the United States Geological Survey (USGS), 2004 was the second deadliest year for earthquakes in recorded history, an unenviable statistic that was dominated by a single event—the Asian earthquake and, in particular, the Indian Ocean tsunami that followed it. But how does 2004 stand in the vast geological timescale?

Environmental change brought about by natural disasters has

223

occurred throughout the history of the earth, and the consequences have sometimes led to large-scale bio-extinctions. For example, volcanic activity associated with the Siberian Flats is thought to have played a major part in the *Great Dying*, the most severe species extinction in the history of the earth, when approximately 70 percent of all terrestrial species were lost (see Chapter 13).

Large-scale volcanic events like the Siberian Flats are rare, and only eight have occurred during the past 250 million years. So what can we expect in the next few hundred years? Certainly, more of the same—by and large, volcanoes, earthquakes, and tsunamis of magnitudes we have become accustomed to. These natural disasters will continue to occur, and although they will always bring death and destruction we can, to a point, forewarn ourselves of their coming and mitigate against their effects. Or, at worst, we can recover from them.

MEGADISASTERS

But what of megaevents that bring disaster on a global scale and that might make full recovery impossible? Are they likely? The answer in the long term, on the geological timescale, is probably yes. As the plates of the planet's surface continue to move and collide, megadisasters will go on occurring.

Some scientists believe that disaster events that can have a massive environmental effect may even now be lurking in the wings. These events include the *supervolcano* and the *megatsunami*.

▲

Supervolcanoes are formed when magma from the mantle moves upward but fails to reach the surface. Instead, it fills a very large chamber beneath the crust until a tremendous pressure begins to build up. The pressure continues to increase as volcanic gases are trapped within the magma, and the process can go on for thousands of years until a supervolcano finally erupts in a destructive explosion that is many, many times greater than any "normal" volcanic eruption.

The last supervolcano eruption was that of Toba, in Sumatra, around 75,000 years ago, which had a volcanic explosivity index of 8. Toba caused an environmental catastrophe; one of the major hazards arose from the volcanic ash and gases thrown into the atmosphere, which created darkness and acid rain. In the ensuing *volcanic winter*, it is thought that at least half of the human population was wiped out.

At present, a potential supervolcano lies under Yellowstone National Park in the western United States. An eruption here would have disastrous consequences not only for the United States, but for the world if another volcanic winter sets in.

▲

A megatsunami is generated when there is a release of energy greater than that required to initiate a normal tsunami. One scenario for the generation of a megatsunami was proposed by Simon Day and Steve Ward, who have studied volcanic activity in the Canary Islands off the coast of North Africa. Cumbre Vieja is an active volcano on the island of La Palma. If it erupts again and collapses, the west flank may fail, giving rise to a landslide that would dump a huge block of rock into the Atlantic Ocean. This event would generate a tsunami perhaps over a hundred meters high that would propagate across the Atlantic and strike, among many other locations, the eastern seaboard of the United States. However, other scientists have raised doubts over whether the wall of the volcano would fail, or even if it did, whether the resulting tsunami would be as big as predicted.

Thus, speculation on the possibilities of great volcanic, earthquake, and tsunami disaster events will no doubt continue to come and go.

INDEX